EVOLUTION GONE WRONG

ALEX BEZZERIDES
with illustrations by Peter Davidson

EVOLUTION GONE WRONG

THE CURIOUS REASONS WHY OUR BODIES WORK (OR DON'T)

Recycling programs for this product may not exist in your area.

ISBN-13: 978-1-335-69005-0

Evolution Gone Wrong: The Curious Reasons Why Our Bodies Work (Or Don't)

This edition published by arrangement with Harlequin Books S.A.

Hanover Square Press
22 Adelaide St. West, 40th Floor
Toronto, Ontario M5H 4E3, Canada
HanoverSqPress.com
BookClubbish.com

Printed in U.S.A.

For Julie and Ellie

Contents

Introduction

It's Not Your Fault

Which animal climbs down from its tree once a week to do its business?

a. koala bear
b. three-toed sloth
c. common ringtail possum
d. spider monkey

I would be willing to bet at least one of these statements is true about you: you had your wisdom teeth pulled, you wear glasses or contacts, you snore, you have sore feet, you have back pain. If you are unlucky, all of those statements might be true. You have probably spent considerable time and money in an effort to alleviate the pain and discomfort from such ailments. From dental work to prescription lenses, pharmaceuticals to customized mattresses, humans spend a lifetime trying to relieve symptoms caused by their anatomical imperfections. But have you ever stopped to consider why those aches and pains occur in the first place? Why don't our teeth fit in our mouths? Why do so many people have blurry vision? Why are our knees and ankles always getting sprained and twisted? Why are our backs so delicate we have to worry every time we bend over to pick something up? It is not just the aged who suffer from these maladies. Seemingly from birth, humans are uniquely prone to bodily discomfort and malfunction.

This book is about the aches and pains of the masses and why they happen. It is not about obscure conditions you only ever hear about from some aunt you see during the holidays. It is about your body and why it often does not work as well as you might hope or expect.

Don't worry, the book is not a long, drawn-out guilt trip; the focus is not on all the problems humans bring upon themselves with their unhealthy behaviors. Thus, topics like atherosclerosis, lung disease, and cavities are off the table. I am not going to give you a hard time about your diet, how much you exercise, or how infrequently you floss. Instead, this book will provide explanations for anatomical imperfections that are *not* your fault, like having crooked teeth or flat feet or waking up with a sore lower back. But why do those issues happen? What happened along the long, complicated arc of human evolution to leave us with bodies that fail us in such predictable ways?

BIOLOGICAL MASH-UP

The causes of our bodily shortcomings are rooted in both evolution and anatomy. I am uniquely qualified to cross over between those two disciplines, kind of like Taylor Swift moving between country and pop. My country music years were my graduate training in evolution and ecology. When not doing research and writing journal articles, I was teaching, and I learned that I preferred being in front of a class to being behind a lab bench. So after finishing graduate school, I applied for teaching-heavy appointments rather than jobs with more of a research focus.

My first academic job was at a small college in northern Wisconsin, where they needed me to teach human anatomy and physiology (A&P). Like Taylor Swift switching to pop, I pivoted and shifted my focus as a biologist from asking questions about natural selection and population ecology to learning the best ways to teach students about digestion, respiration, and the endocrine system. Unlike Taylor Swift, my starting salary was $41,000, which was somewhat humbling having recently finished a decade of undergraduate and graduate studies. It turns out they're right when they say you don't get into teaching for the money.

I eventually ended up at a small college in the western United States. I still teach a significant amount of A&P, but working at a small campus allows me to throw a broad disciplinary net. Over the years I've taught classes that run the gamut from entomology to human dissection. When people ask what type of biologist I am, I never know quite what to say. I am a biologist who can identify beetles, but I can also tell you how your kidneys work. I am not sure what type of biologist that makes me. I think it just makes me a biologist, period.

MY PAIN IS YOUR PAIN

I was not inspired to write this book by some set of unbelievable circumstances. I never lost half of my large intestine in a grizzly bear attack. I did not nearly die from Dengue fever while attempting the first descent down an unexplored river in Africa. I do not suffer from any unusual genetic conditions, autoimmune disorders, or physiological anomalies. I have a normal, middle-aged body, which means I have the same types of problems everyone has. My teeth did not come in particularly straight. I blew out a knee playing basketball. Sometimes my back gets sore enough that it is hard for me to sleep.

I did not think much about these ailments for years because none of them caused me an inordinate amount of trouble. I was also busy learning and teaching about all the ways the body works when it functions "normally." Human anatomy and physiology is a complicated subject, and it takes a long time to get a handle on how the body runs under typical, pain-free circumstances. There are several good A&P textbooks, and all of them are more than 1,000 pages long. In the early years of my career, I was not thinking about the origins of conditions like myopia or bunions or gestational diabetes. I was more concerned with distilling those 1,000-plus pages into two semesters of course content in order to teach scores of future nurses and doctors how eyes, feet, and uteruses typically work. Questions about issues like choking, torn menisci, and the pain of childbirth came to me later after I had established a foundation of knowledge about how our bodies work.

"HOW" QUESTIONS

Biologists, and their students, learn about bodily systems by asking questions. They can ask proximate questions in which they try to understand *how* something works. They can also ask ultimate questions where they try to answer *why* a structure, process, or behavior is the way it is. Proximate questions take up most of the time in an A&P class. How do heart valves control blood flow? How does the ACL restrict the movement of the knee? How does the cornea refract light? Most of what we know about human anatomy and physiology comes from answers to those types of "how" questions. It makes sense because if you need to repair a heart valve, put a knee back together, or use a laser to correct someone's vision, you need to know how the structures work in order to fix them.

"How" questions can have tremendously complicated answers. Teaching and learning A&P often does not extend beyond the proximate level because it takes so long to understand proximate-level questions and answers. For example, to teach just the basics of muscle contraction it takes several drawings, considerable discussion, and more than an hour of class time to cover physiology that, in real time, occurs in a fraction of a second. And, mind you, this happens in a lower-level undergraduate class. If students revisit the subject in graduate or medical school, they learn many more layers of detail.

"WHY" QUESTIONS

Even while spending so much time on proximate mechanisms in A&P, I never completely broke the habit of asking ultimate questions because of the other classes I teach. Ultimate questions get at the evolutionary roots of a topic. For example, in my class on animal behavior, instead of teaching *how* dolphins jump out of the water (which involves much of the same minutiae about nerve conduction and muscle contraction covered in A&P), I ask my students to think about *why* dolphins jump out of the water. "Why" questions make for great exercises in hypothesis generation and experimental design. Do dolphins jump to avoid predators? Do they jump to communicate messages to each other? Do they jump because it is a more efficient way to travel? Do they jump because it's fun? "Why" questions are easy to ask but notoriously difficult to answer. It is one thing to brainstorm about why dolphins jump. It is an entirely different exercise to design and conduct experiments in an attempt to answer the research question. As evidence of the difficulty, scientists have yet to settle on an unequivocal answer to the dolphin-jumping question.

Another example I use in my animal behavior course to illustrate ultimate lines of questioning introduces students to the unusual bathroom behavior of sloths. Three-toed sloths spend nearly their entire lives in the forest canopy. The ground is the most dangerous place for sloths because there are hungry, toothy predators down there like jaguars and feral dogs. Once a week, however, sloths break their arboreal habits and head to the ground to take care of some personal business. They slowly climb down to the forest floor to defecate. (Don't worry if you missed that first multiple choice question. You'll have plenty of chances to redeem yourself.)

On the surface, this behavior is baffling. Why risk the chance of encountering a predator? Why not just let it fly from the branches? In class, my students work together to develop hypotheses and design hypothetical experiments to test their hypotheses. Are sloths fertilizing their trees in a targeted manner? Is it some way of marking their territory? Is it an atypical type of mate attraction?

Acutely observant scientists solved the mystery only recently with a great deal of patience. They first observed that sloths have algae growing in their fur, which gives the sloths a green tint. The algae help the sloths blend in with the forest canopy, but the story goes beyond organic camouflage. The sloth scientists noted sloths feeding on their homegrown algae and in doing so, supplementing their otherwise nutrient-poor diet. Eating their own fur algae is admittedly weird, but it gets even stranger than that. A population of moths lives in the fur of each three-toed sloth. The moth population increases the nitrogen content of the fur and thus promotes the growth of the algae the sloths snack on. When the sloths make their weekly treks to the bottoms of trees, the female moths lay their eggs in the fresh sloth dung. The tidy sloths cover up their mess with some leaf litter, and after the eggs hatch, the moth caterpillars dine on the sloth poop, grow up, become adults, and fly to the canopy layer to colonize sloths just as their parents did.[1]

Why do three-toed sloths come down from their trees to defecate? Sloths risk their lives to make a dung nursery for the moths on whom they depend for fertilizer to grow the algae they not only use as camo but also eat from their own fur for an extra shot of nutrition. Bam! Mystery solved. We can finally let the sloths poop in peace. Next question.

I hope this sloth-and-moth story has made the point that ultimate questions are fascinating to consider. They push researchers in completely different directions compared with proximate questions. The answers to ultimate questions are also often wildly unexpected. It is a crucial point to make because, more than anything else, this book is about ultimate questions. I'm just going to focus on ones about the complicated evolution of the human body instead of ones about pooping sloths.

DIGGING DEEP

After years of thinking about "why" questions in the context of dolphins, sloths, and other nonhuman animals, I started asking "why" questions about human anatomy. Why are humans prone to choking? Why is infertility such a widespread problem for couples? Why do females menstruate? The first such "why" question hit me one day when discussing teeth in A&P. I had each of the students raise one hand and then asked them to lower their hands if they had worn braces. A good portion of the class lowered their hands. I asked the remaining students to lower their hands if they had had any wisdom teeth pulled. Now only a few students' hands remained in the air. Finally, I asked them to lower their hands if their teeth were crooked. We started with more than 70 hands in the air and ended with only two. Out of 70, only two students' teeth naturally fit well in their mouths.

I could not stop thinking about all those braces, pulled wisdom teeth, and crooked smiles. Why are human teeth such a poor fit for the jaw? Many more questions quickly came to mind about the imperfections of eyes, throats, knees, feet, backs, and more. Why are the esophagus and trachea next to each other, leading to choking issues? Why do so many people tear their menisci and ACLs? Why is breastfeeding so routinely painful for humans? Once you start asking these "why" questions, they can spiral out of control like a kid asking why, why, why?

I needed some answers. I started to read everything I could find about anatomical shortcomings and the evolution of the human body. After wading through hundreds of studies published in an incredibly wide array of journals, I have learned the answers are often as unpredictable as those discovered in the sloth-and-moth story. To answer questions about human bodies, I read about slimy hagfish, gorilla penises, and chimpanzees running on treadmills. The answers extend beyond anatomy and into behavior. Thus, I also learned about the origins of bipedalism, running, and hunting in early humans and exactly how long the carcass of a boar can sit out before it spoils. Spoiler alert—it's not very long.

I say we get on with it. These questions are not going to answer themselves, and I imagine you are as curious as I am about the anatomical failings of the human body—your body. So put on your glasses, slip a pillow behind your back, and make yourself comfortable. Why is the human body so uniquely prone to aches and pains? How did evolution lead us so astray? Turn the page and we'll see if we can find some answers.

PART ONE

It's All in Your Head

1

Mastodon Stew

What is the record number of teeth pulled from a human's mouth?

a. 24

b. 48

c. 232

d. 584

It feels like I've gone back to middle school. I finally ran out of excuses, swallowed my pride, and had braces put on my crooked teeth. I almost ended up with naturally straight teeth. I emerged from the awkward teenage years with my pearly whites nicely tucked in rows, but then, sometime during college, all my wisdom teeth erupted and shoved things around a bit. I blissfully ignored the issue for a couple of decades but eventually reached the point where I could no longer nip through a piece of lettuce with my crooked incisors. I did not want to gum leafy greens for the rest of my life, so I bit the bullet and made an appointment with an orthodontist.

There are entire medical disciplines founded on the anatomical premise that our teeth, left to their own devices, will eventually end up looking like a windblown picket fence, skewed and overlapping. Every anatomy text on earth has an illustration of the typical human mouth, and it contains exactly 32 superbly straight teeth. Yet in the real world, the vast majority of adults fall short of this picture of dental perfection. Most people either need braces, have some wisdom teeth pulled, or just tolerate having crooked teeth.

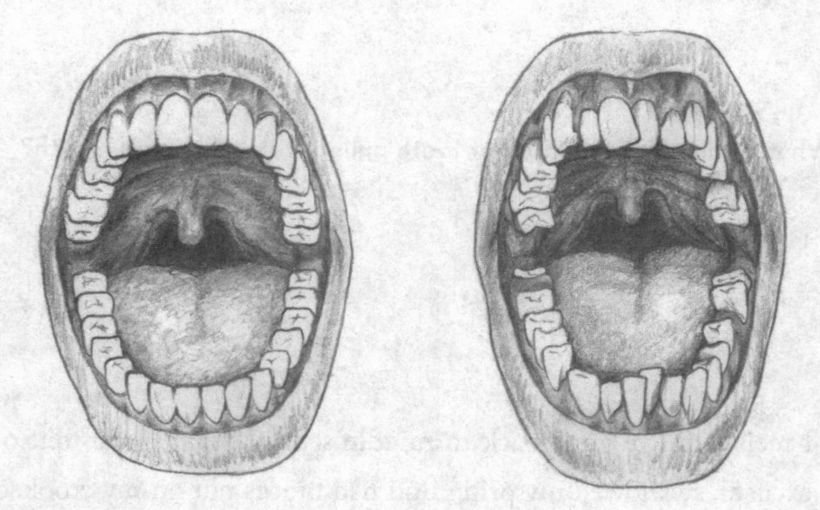

After teaching anatomy to undergraduates for several years, it really started to bother me that I couldn't explain to students *why* our teeth don't fit in our mouths. The students would learn the basics of dental anatomy, recognize the existence of the tooth–jaw mismatch, and then we would move on to the next feature of the digestive system. But it started to keep me up at night. Why don't our teeth fit in our mouths? Braces and pulled wisdom teeth are so common we don't stop to ask the question. What if nearly everyone had their pinky toe amputated during adolescence to improve their gait? What if earlobe removal was a routine, medically necessary procedure? If aliens came down and saw us removing our own teeth, it would seem as strange to them as those examples.

I don't like losing sleep, so I started reading about the issue. A lot. It turns out "why don't our teeth fit?" is a very simple question with a fascinating, complicated answer.

WHICH CAME FIRST, JAWS OR TEETH?

Everyone has heard the age-old expression "You are what you eat." In many ways, "You are what you ate" is a more accurate saying. This version of the popular aphorism is not talking about last night's fried chicken or this morning's toast and coffee. Rather, it hearkens back to our long-distant ancestors scraping out a living eating tubers, bulbs, and roots millions of years ago on the African savannah. It also refers to our more recent ancestors who chased down large animals, brought them back to primitive camps, and shared them around the fire.

The "You are what you ate" version is nowhere more true in the body than in the shape and structure of the human jaw and teeth. Even though the type and abundance of food we rely on as humans has changed rapidly in the past, modifications to hard and fast structures like molars and mandibles are very slow-moving processes. The teeth and bones that make up the jaw can, and do, change over time, but they are not able to change quickly enough to keep up with the moving target that is the human diet. The result is that throughout history the human mouth has had to do work for which it is not particularly well suited.

A fundamental question should start this discussion: which appeared first, jaws or teeth? The answer is less obvious than with the chicken-or-the-egg question everyone likes to ask.* The confusing thing about the jaws or teeth question is that throughout the evolutionary history of animals, hard features called odontodes have ended up in all variety of places other than the mouth.[1, 2] Odontodes are so hard they are referred to as tooth-like structures. It's kind of weird to think of teeth outside of the mouth, but we do live in a world in which crickets have ears on their legs and kangaroos have three vaginas, so, really, it's not all that odd. Odontodes fossilize like a million bucks (being, you know, very hard), and, therefore, there exists a rather complete record about odontode evolution going back all the way to the earliest vertebrates. Long story short, it turns out there were teeth (or at least tooth-like structures) quite a long time before there were jaws. They just were not all nicely organized within the oral cavity as one typically imagines them.

Those odontodes were either flat structures called dermal denti-

---

*With the chicken-or-the-egg question, all evidence points to the presence of eggs long before there were ever chickens. Birds are just flying reptiles, and there were many, many millions of years of egg-laying, non-avian reptiles before anything like a fancy, feathered chicken ever evolved.

cles that toughened up the skin of ancient fish or they were pharyngeal teeth down in the throats of primitive ocean dwellers. Denticles survived all the way to the present time and are still observable as the scales of cartilaginous fishes including sharks, skates, and rays. One also need not look very far for evidence of pharyngeal teeth in extant (living) organisms. In fact, there are pharyngeal teeth in millions of homes throughout the world—the ubiquitous goldfish is a nice example of a modern organism with its teeth in its throat.

The real debate here is from which set of these primitive teeth (the skin ones or the throat ones) did the teeth in the oral cavity originate? Baring their teeth in one corner, we have the proponents of the outside-in hypothesis. As the name suggests, this camp makes the argument that the dermal denticles, or skin teeth, eventually migrated around and into the oral cavity. In contrast, the inside-out hypothesis backers make the case that teeth arose in the pharyngeal cavity and eventually migrated up to the oral cavity.

Regardless of whether they originated in the skin or in the throat, teeth eventually ended up in the mouth. Jawless fish with oral teeth called agnathans are well represented in the fossil record. The now extinct eel-like conodonts swam in ocean waters starting around the dawn of the Paleozoic era (542 million years ago) and had a good, long stretch of success before running into trouble a few hundred million years later.

Being a jawless fish with teeth works quite well in a world filled only with other jawless fish, trilobites, and squishy marine worms. It's even manageable in a world where fish have jaws but no oral teeth.[3] It works significantly less well in a world filled with sharks.

Once the Devonian period of the Paleozoic era came around (the Age of Fishes some 400 million years ago), life became much more of a struggle for the agnathans. Most of them did not survive the carnage of the Devonian. Around 120 species of jawless hagfish and lampreys represent the few stubborn and persistent agnathans that made it through to the present day.

It is rough evolutionary sledding for the average transitional organism like a hagfish. Fancier, craftier, more highly evolved life forms typically spell the end for transitional species. Hagfish managed to buck this trend. Seeing one up close gives some indication as to why they made it through the evolutionary gauntlet. They are one of the ugliest, slimiest, least desirable-looking animals to ever swim the ocean blue. They may have survived, in spite of their nonexistent jaws, precisely because they are terrifically unappetizing.

Hagfish—the ultimate evolutionary survivors!

In an unfortunate recent twist of fate, the hagfish has not proven itself unappetizing enough for humans, with our rapacious appetite for all things exotic. They are considered a delicacy in some cultures and are also used in the production of leather goods. Apparently, there are people, somewhere, who cannot bear to live without hagfish leather boots. As a result, many hagfish species are now endangered because of overfishing.[4]

As much as I like to think about the outside-in and inside-out scientists coming to blows at cocktail parties or conferences focused on the evolution of vertebrate dentition, what matters here is to understand that jaws and teeth have independent evolutionary origins. Recognizing the evolutionary independence of jaws and teeth is critical to answering the "why don't our teeth fit?" question. When one of the structures underwent a notable change (like the jaw becoming smaller, for example), the teeth did not automatically follow suit. A mutation in a gene that controls the size of the jaw cannot magically spur a mutation in a gene controlling the size of the teeth. If cells could presciently tie mutations together in such a way, we would all have perfect teeth and orthodontists would have to find a new gig.

FROM FISH TO REPTILIAN TO MAMMALIAN TEETH

With the foundation of jaws and teeth established, it is time to take this story out of the water. After all, the ultimate goal of this first chapter is to understand why so many of us needed braces in the seventh grade, and seventh grade decidedly took place on land.

Around the time the ocean started to fill up with toothy and jawed beasts, some proactive ocean dwellers took a crack at living on terra firma. Of course, this happened over countless generations and we'll probably never know if it was to avoid being shark bait, to find some better foraging opportunities, or for reasons we have not even considered. Regardless, some animals that spent much of their existence in the shallows started to spend some of their time on the shore. This transition has been well chronicled in recent years with the discovery of incredible transitional fossils like the prehistoric fish *Tiktaalik*.*

Tiktaalik lived north of what we now call the Arctic Circle, and it had unique skeletal features for a fish, like a neck and a flat head. It could bear weight on its fins and push off the bottom of the shallow waters where it lived. *Tiktaalik* was, effectively, a bridge between fish and amphibians, exhibiting both fish-like and amphibian-like features.

Tiktaalik crawled around some 375 million years ago, leaving about 150 million years until the appearance of the first mammals on Earth, somewhere in the Triassic period. Obviously, some significant changes occurred to terrestrial vertebrates during those 150 million years. The skin of amphibians hardened, allowing for better water conservation, and vertebrates started to spend longer and longer stretches away from water. The limbs became modified for land travel, and reptiles picked up speed and skill as predators.

---

*For a thorough description of *Tiktaalik* and the events surrounding its discovery, read Shubin, Neil, *Your Inner Fish*. New York: Vintage Books, 2009.

Anatomical features near and dear to our hearts, like nipples and hair, pushed a few select animals past the reptilian stage and into the dawn of mammals. Most early mammals were not the most impressive lot; imagine little squirrel-like animals dashing through the legs of much larger dinosaurs and you've got the idea.

Significant changes occurred during this time, altering jaws and teeth. As reptiles radiated out to fill the wide variety of terrestrial environments, their teeth and jaws became highly varied. Some, like turtles, bailed on teeth entirely and adopted the tough-as-nails beak variety of feeding apparatus. Many, like the bulk of lizards, ended up with teeth attached to the inner side of the jaw, as opposed to teeth stuck down in a socket like those that appear in a few other reptiles (e.g., crocodiles) and in mammals.

Other differences beyond attachment style arose between mammalian teeth and the earlier teeth of fish and reptiles. The most obvious was a significant reduction in the total number of teeth moving from fish to amphibians to reptiles and on to mammals. The average run-of-the-mill reptile might have as many as 200 or 300 teeth, while it is unusual for a mammal to present with more than 50.* It is rare, but on occasion, a human will develop significantly more teeth than the typical number of 32. Do you remember the question from the start of the chapter? It asked about the record number of teeth pulled from someone's mouth. I wouldn't judge you for answering 24, since adults should have, at most, 32 teeth.

---

*The Virginia opossum tops the list of North American mammals, with 50 teeth. It is also the only North American marsupial found north of Mexico.

An extreme case occurred in 2014, however, when a teenager in India had 232 teeth pulled, leaving him with 28, the typical number of teeth for an adult with their wisdom teeth removed.

Owners of cold-blooded pets may also notice that, unlike mammals, fish and reptiles continue to replace their teeth throughout their lifetimes. Mammals lose their primary teeth relatively early and then must bear the responsibility of maintaining their second set for the remainder of their lives. This becomes a serious issue as mammals age; the lack of a sufficient number of healthy and strong teeth can spell their demise. The issue is not an insignificant one for humans. People in the United States in their 20s and early 30s have an average of 26.9 teeth, but that number dips to 22.3 for people 50 to 64 years old.* Maintaining an adequate number of healthy teeth is critical, with one study demonstrating a 30% increased risk of death for elderly people with fewer than 20 teeth compared with those with fuller sets in their mouths.[5]

The other significant difference one would note upon simultaneously staring down the gullet of a lizard and a house cat is a wide variety of tooth types in the cat, whereas the lizard teeth all basically look the same.

Heterodont dentition, meaning having different types of teeth, is a hallmark of mammals—it allows mammals to eat a diet unparalleled in its diversity. This type of dentition has secondarily

---

*The data about tooth loss are available online on a page dedicated to research from the National Institute of Dental and Craniofacial Research, which is under the broader umbrella of the National Institutes of Health. Google NIH tooth loss overview to find the report.

Most reptiles have large numbers of similar teeth. Mammals have fewer teeth, but they come in different shapes and sizes.

evolved in a select few other groups, such as in crocodiles or in the highly specialized fangs of some types of snakes. Heterodont teeth are what allow us to tear a piece of meat off a chicken bone with one bite, crunch through a carrot with the next, and slowly savor a chewy piece of caramel for dessert. Such dietary versatility has allowed mammals to conquer all corners of the globe because no matter where we go, we can usually find something to eat.

PRIMATE TEETH

And then, around 66 million years ago, a giant asteroid came along. When I say giant, I mean six-miles-in-diameter giant. A six-mile-diameter space rock striking the earth did not have a positive impact on air quality. The mass extinction event that occurred at this time is known as the K–Pg boundary or the K–Pg extinction event (for defining the border between the Cretaceous and Paleogene time periods).* Multiple lines of evidence suggest that a giant asteroid striking the Yucatan Peninsula in present-day Mexico caused the calamity.

The result was the extinction of 70 to 80% of organisms on Earth. Although this was clearly a negative event for the bulk of animals, the outcome for mammals, who were exclusively nocturnal at this point in time, was not necessarily terrible. With most of the pesky dinosaurs out of the picture, the remaining mammals could finally come back out to play during the day, and they began to thrive.

With mammalian evolution no longer constrained by dinosaurs, mammals radiated out in all variety of manners in the same way the reptilian lineage had expanded 200 million years earlier. The original formula for mammalian permanent tooth dentition consisted of 11 teeth in each quadrant for a total of 44. Eventually, as mammals spread out to fill nearly all the diverse habitats on Earth, the number and shape of teeth began to vary based on the dental needs in those different environments. A field mouse

---

*You may have heard of the event as the K–T boundary. K–Pg has replaced K–T because geologists have soured on using *Tertiary* as a formal distinction. *K* is used instead of a *C* to avoid confusion with the Cambrian or Carboniferous eras, which are sometimes abbreviated with the letter *C*. *K* was selected because the Cretaceous rocks are often chalky in composition, and both the Greek and German words for *chalk* start with *K* (*kreide* and *krete*).

scraping out a living eating seeds and small insects on the African plains, after all, has distinctly different demands on its teeth than a saber-toothed cat dragging down and eating a rhinoceros.

A few mammals ascended into the trees and eventually became primates. The arboreal diet was different from the terrestrial diet, and primate teeth began to diverge from those of other mammals. One primate lineage, the new-world monkeys (golden lion tamarins and howler monkeys are nice examples), ditched one of the incisors and one of the premolars from each quadrant to cut down the number of teeth to 36. The other primate lineage, the old-world monkeys (like baboons and macaques) and the apes, also shed an incisor. In addition, they lost two of the premolars, leaving the standard adult complement of 32 teeth.[6]

Thirty-two became the set number of teeth for our lineage somewhere around 30 million years ago. The last 30 million years seem like a sufficient amount of time for our jaws and teeth to sort out their differences. There were, however, other big changes to come that dramatically altered the shape of our ancestors' mouths.

GETTING OUT OF THE TREES

Life hummed along relatively uneventfully for primates for the next few million years. No fabulously horrific extinction events or other earth-shattering phenomena dramatically changed the course of life on our warm, little planet. Primates continued to put their snazzy thumbs to work and carved out a nice niche for themselves, living predominantly in and among trees and enjoying a diet of fruit and leaves. The group continued to radiate, and the great-ape branch started to take on definition. Many of the changes from ape to great ape were skeletal in nature. The shoulders and rib cage started to take on a more human-like appearance, and the spine and hips began to shift, allowing for more upright posture.[7]

Eventually the great-ape branch of the evolutionary tree also diversified and split. Researchers continue to debate the date of the last common ancestor of humans and chimpanzees, with estimates ranging from 5 to 13 million years ago. Several differences come to mind when thinking about humans and our closest living relatives, the chimpanzees, and many of them are either indirectly, or directly, linked to our poorly fitting human teeth.

The first notable difference is that most of us did not wake up in a tree this morning. Recent fossil evidence suggests human ancestors bailed on the arboreal life somewhere around 3 to 4 million years ago.[8] There is general agreement that Earth's climate underwent a significant change around this time. A dramatic change in climate would have notably altered food availability. Before such a shift, with adequate food, shelter, and safety, there would have been no motivation to upset the apple cart and try out a totally different means of existence. Leaving the trees was not a step our ancestors chose to take as much as one they *had* to take. Those individuals who could scratch out a living outside of the trees survived, reproduced, and passed on their genes. Slowly but surely, with slight modification built upon slight modification, hominins (the group that includes humans and our extinct close relatives and ancestors but not the other great apes) adapted to life on the ground.

The dietary changes from a life in the trees to a life on the ground had a significant effect on teeth and jaws. Being mammals, human ancestors had a diversity of tooth types, which made it possible for them to tolerate a dramatic shift in diet. Some very clever studies by a group at the University of Utah demonstrated the change in diet by analyzing fossilized teeth of human ancestors from around this time.[9]

Sparing the gory details of radioisotope biochemistry, what we eat leaves a molecular footprint throughout our bodies. Different types of plants leave different types of footprints, and by analyzing tissue (in this case tooth enamel), it is possible to get a sense of what individuals ate. Starting around 4 million years ago, the hominin diet transitioned from fruits and leaves to grasses and sedges. The demands on teeth were quite different, naturally, after the switch from munching a juicy mango to chewing wiry grasses. Just as there is with human teeth today, there was considerable variation in hominin teeth, and those individuals with teeth appropriately sized and shaped for the new grassy diet would have been at a distinct advantage. The tooth-lottery winners of the time passed their lucky teeth genes on to their offspring, and the nature of hominin teeth changed. One of the most dramatic changes was to the size of the molars. In short, they became huge.

Some early hominins, like the famous *Australopithecus afarensis* fossil Lucy, had giant molars. Another major change distinguishing us from chimpanzees was that hominins started to move around on two feet. Effectively, at this point in time, and for a couple of million years, our ancestors became bipedal cows.

Those giant molars are at the root of the modern problem. They were not an issue way back then because, at the time, there was also a large, powerful jaw to house those large, powerful molars. Tooth size was a decent match with jaw size for a hominin like Lucy. I have never seen it suggested, but I would wager she had lovely teeth. Unfortunately for the tooth–jaw marriage, the grass and sedge diet did not last forever. A diet of grasses and sedges was perfectly acceptable when hominins did not have to travel great lengths to find them. As a changing climate increased the travel distances required to obtain food, there was pressure to find something more calorically rich to meet the energetic demands of foraging. Something with a greater bang for your buck. Something like meat.

THE MEAT SWEATS

The problem with meat is that it does not want to be eaten. One way around this complication is to let someone else do the killing and then feed on the leftover scraps. Many paleoanthropologists now argue this was the solution employed initially by hominins. Even scavenging carrion presented some new challenges for the hominin line, however. The hominin jaw and teeth were well set up for eating salad, not for tearing muscle away from bone. Primitive tools were employed to help with the challenge.

Around 2 to 3 million years ago, our ancestors started using sharp rocks to hack pieces of meat off bone. Whether or not they were fashioning the sharp rocks or simply finding them ready to use is a matter of debate. Just like a life of running around and eating grass was not sustainable for the hominin line, a life of subsisting on raw carrion was also destined to be short-lived. If significant quantities of meat were going to be part of the hominin diet, eventually something other than leftover, already-dead scraps had to be on the menu. And while knocking a sharp rock on a scavenged meat-covered bone does not necessarily require a great leap forward in intelligence,[10] actually figuring out how to kill a decent-sized animal requires significant levels of brainpower.

The dietary transition from carrion to fresh meat would not have been trivial for our ancestors. Most predators rely on a combination of guile and speed to bring down their prey. Early hominins had guile in spades, but they didn't have the quickness of most quadrupedal hunters. Bipedalism is highly advantageous when traveling long distances, but over short distances it is a slow way of getting around.

An Olympic champion sprinter might seem fast when compared with other humans, but my fat, old, lazy housecat, Tyson, could still beat a gold medalist in a short race.* Without great bursts of speed, to hunt effectively, humans had to learn to outwit their prey. And even if they could chase after, catch, and wrestle

---

*Olympic sprinters top out at around 45 kph (28 mph), while housecats can clock in at 48 kph (30 mph). Scale up the size of the kitty to a cheetah and the comparison becomes ridiculous, with cheetahs hitting speeds north of 112 kph (70 mph).

a gazelle to the ground, what were they going to do then? Attack it with their dainty nails? Try to tear into it with their pathetic canines? At best, early hominins trying to hunt like a lion would end up exhausted and dirty. More likely they would end up gored and bloody.

Hunting as a soft and slow human demands intelligence. The smartest hominins were the ones who could work rocks into sharp spears and cooperate to use tools to bring down big game. Killing large prey would also have required significant social interaction, another trait typically associated with increased cranial capacity. Reciprocity would have played an important role in our ancestors' lives. Hominins and early members of *Homo sapiens* must have had both good and bad days at the office.

A good day meant plenty of meat to share. On bad days, their spears would have missed their marks, and early hunters would have slinked back into camp with nothing more than a couple of nuts and berries picked up along the way. On those days, the empty-handed hunters needed some help from their fellow cave dwellers. Handouts would most likely have come from individuals who either had benefited from sharing in the past or who anticipated reciprocity down the road. Hominins were less likely to go to bed hungry if they could effectively navigate interpersonal relationships.

The benefits of cooperation, sharing, and communication likely played a significant role in driving the increase in brain size. As members of hominin societies grew smarter and smarter, they would have, in turn, become increasingly successful hunters. Our ancestors would have entered a positive feedback loop where the smarter they got, the more successfully they could hunt. The more successfully they could hunt, the smarter they got by feeding a growing brain in demand of energy. The intelligence race was on, and around this time, the first members of our genus *Homo* appeared, and the average cranial capacity increased from around 400 cm^3 to 650 cm^3.

The selection for giant noggins rolled on in the ensuing 2 to 3 million years. We weren't satisfied with a slightly larger brain. It has ballooned to an average volume of 1,350 cm^3. Human brains grew so big that they are now, finally, held somewhat in check by natural selection. The check for the human brain is, of course, birthing the swollen thing. The brain has gotten just about as big as it possibly can while still allowing it to get out of the mother's body. It has pushed the envelope so much that sometimes it cannot even come out, but that's a story for later.

At the same time selection for a giant brain was increasing, the selection for a ridiculously powerful jaw was decreasing. As the diet changed, the need for those giant molars and giant jaws diminished. With the selective pressure for a powerful jaw reduced, it was possible for mutations weakening the jaw to persist within the hominin genome. Around 2.4 million years ago, such a mutation took place.[11] A variation popped up in the hominin lineage that significantly weakened the jaw. It is entirely possible similar changes affecting the structure of the bones and muscles of the jaw had appeared before, but if the timing were off, those changes would have faded away. If the mutation came along during a period when hominins needed big, powerful jaws, it would not have persisted. Chimpanzees still need large, robust jaws, and as such, their jaws have not shrunk, as human jaws have, over the last few million years. Eventually, the timing was right for a jaw-weakening mutation to stick, and the human jaw began the great shrink to its current weaker state.

GROUNDED FOR LIFE

Another leap forward was necessary to fully leave behind the raw-carrion lifestyle. Hunting had solved the carrion part of the problem, but the meat was still raw. The only path away from the raw-food diet involved learning to control fire. As with hunting, however, controlling fire takes some serious wits.

Many modern humans struggle to pull it off even when provided dry newspaper, perfectly split wood, and a box of matches. It is not a skill that comes to us naturally. I will never forget the first time my brother, Nick, and I tried to make fire. I was around four years old and he was six or seven. We set up shop in the garage, hoping our parents wouldn't find us. My job was to steal matches from the house while Nick organized our combustibles. We had some kindling and two-by-fours, but to our great disappointment, we never got much of anything aflame. After several failed attempts, we were busted when our dad opened the garage door—and promptly grounded us for eternity.

Some have suggested that the first fires hominins controlled were ones they poached from nearby lightning strikes. Even if this were true, it still would have taken a great degree of wherewithal to keep the flame going. Eventually, it would have gone out ("Who was watching the fire?" "I thought *you* were watching the fire?!"), and the only sustainable solution would have been the ability to light it again without having to wait for lightning. A damp day would have greatly tested the increased cranial capacity and processing power of early hominins as they tried to make fire.

Paleoanthropologists have hotly debated the origins of hominin-controlled fire for years. Some place the date as far back as a million years,[12] others put it at more like 350,000 years ago,[13] and still others argue this crucial milestone occurred as recently as 12,000 years ago.[14] What is *not* in dispute is that at some point a clever cave dweller figured out how to create a fire, and hominin life has not been the same ever since. Some experts have gone so far as to suggest that our ability to control fire is the very characteristic that makes us human.

THE FIRST BARBEQUE

Eventually, an early human put two and two together and brought a piece of meat over to the sizzling flames. Or maybe the first grilled steak was purely an accident, the result of an out-of-control fire searing a carcass stashed a little too close to the flames.

Regardless of how meat ended up over the flame, it was a fortuitous step for humans. The combination of hunting and the control of fire is at the heart of the story of why so many people needed braces in middle school. Chewing a nicely seared steak is a lot less work than grinding your way through a raw piece of meat. This is especially true if you don't have a knife. Although humans were, by this point, using sharpened rocks to cut meat away from bone, they were not exactly using fancy chef's cutlery. Cooked, tender meat falling off the bone would have made mealtime far easier.

There were several other advantages to cooked meat. Even before hominins started harvesting fresh game, fire would have helped take the funk out of raw carrion. With the advent of scavenging by hominins, prior to fire, bacterial infection had to be a serious concern. Everything about the odor given off by a dead animal suggests that if it has to be eaten, it should be well roasted.*

Even with our modern conveniences for food preservation, humans still manage to get sick quite frequently from spoiled meat. Studies by anthropologists at Harvard have examined how cooking reduces the pathogen load in raw meat.[15] Researchers took the carcass of a freshly killed wild boar and left it out in the open but kept scavengers at bay. At both 12 hours and 24 hours, they took samples of the carcass to see how funky the meat had become. After 12 hours, levels of nasty bacteria like *Escherichia*

---

*Years after our failed fire experimentation, a couple of times during college Nick scavenged roadkill deer as a quick and economical way of acquiring meat. The idea isn't all that crazy; many states have systems in place by which they grant roadkill salvage permits to interested parties. I usually ask for my steak to be prepared rare, but if roadkill tenderloin is on the menu, I ask the chef (my brother) to shoot for somewhere between well done and burnt to a crisp.

coli and *Staphylococcus* had already started to rise. By 24 hours, the levels were downright dangerous. When the researchers roasted the noxious piece of meat over an open flame, they discovered bacteria levels fell by as much as 88%. That might not sound good enough to make it onto the modern dinner table, but for early hominins, it certainly beat rotten, raw wild boar with a side of diarrhea-inducing, potentially lethal bacteria.

There were other advantages to cooking over the open flame. Cooking meat before eating it starts the process of breaking down protein and makes the meat's nutrients more readily available.[16] The growing hominin brain needed all the nutrients it could get. The brain accounts for about 2% of human body mass but uses up to 20% of our caloric intake. By unlocking the true nutritive potential in meat via roasting, early hominins were able to feed their growing brains.

There is one last aspect of cooked meat we should not overlook. It is quite tasty. Simply the thought of a perfectly roasted piece of meat with delicious fat dripping off it is enough to get the digestive juices flowing. We may never pinpoint how, and exactly when, hominins stumbled across the idea of roasting meat, but we can imagine what a pleasant surprise it must have been.

Humans are not alone in enjoying the savory sensation created by roasting. Given the option, all of the other great apes (chimpanzees, bonobos, gorillas, and orangutans) prefer cooked meat over raw.[17] Unfortunately for them, with their comparatively smaller brains, they are unlikely to figure out how to rub sticks together to make fire anytime soon. They might as well be sitting in a garage trying to light a two-by-four by holding a match up to it.

SET IT AND FORGET IT

I like to think about what kinds of meals our ancestors might have sat around and shared. Here is what I imagine a recipe for a meal like mastodon stew would have looked like:

INGREDIENTS: One mastodon, plants, water

DIRECTIONS: Kill, skin, and butcher one mastodon. Cut the mastodon meat into small pieces. Place the pieces into 250 medium-sized pots or one very large pot. Add anything else edible to the pot. Add water. Cook over fire for several hours until done. Feeds 1,000 to 2,000 people, depending on the size of the mastodon.

At this point in the retelling of our history, humans were getting closer to being able to make a meal like mastodon stew. They had the skills and tools necessary to hunt and butcher a large animal like a mastodon. They had control of fire, allowing them to prepare the mastodon in a safe and nutritious way. They needed just one more thing to be able to make a stew—they needed a pot.

Pots are central to the whole stew concept. Compared with spears and fire, pots came along quite late in the tale of human development. Because it was more recent, we have a better idea of when it happened. The most convincing evidence suggests humans first made pots around 20,000 years ago in modern-day China.[18] Having a pot meant food could be cooked for hours until it was so soft that chewing was hardly necessary.

There were other advantages to cooking in a pot. Prior to the pot, all of the tasty, juicy, calorie-rich drippings had been falling into the fire and sizzling away. Now they fell into the pot and became part of the stew. Having a pot also meant spending significantly less time tending to the meal. Humans were able to, for the first time, set it and forget it. Cooking a meal without having to stand next to it the whole time meant more time for gathering food or child-rearing or using one's increased mental capacity to come up with other creative ways of improving the human condition. Having free time is an integral part of the human story, and with increased efficiency of both hunting and cooking, humans were finally able to devote meaningful quantities of time to other tasks.

Around the same time as the creation of the first pot, a particularly clever group of humans in the Middle East hit upon the idea of farming.[19] Combined with the introduction of pottery, farming added a side dish of gruel to the main course of the day. Even individuals with the weakest, smallest jaws, unable to chew meat, could eat enough porridge to acquire the necessary calories to survive. Before gruel, the loss of a considerable number of teeth would have been a death sentence. With mush to eat, there were new options for fussy infants, and humans could keep smiling, toothless grandparents alive for the first time.

FIGHTING THE MISMATCH

The combination of hunting, cooking, and farming created a diet that greatly deemphasized the need for a large jaw and left humans with a mouth full of large teeth that are total overkill for their modern job. In spite of their large brains and relatively small jaws, however, the teeth of hunter-gatherers were still a good fit in their mouths as recently as 10,000 to 15,000 years ago.[20]

There were several reasons for those early, relatively straight smiles. For starters, people 10,000 years ago still had decent-sized jaws, larger than those we have today. Although food availability and processing technology had improved, humans still needed a strong jaw to tear through the meat and vegetables that were the mainstay of the hunter-gatherer diet at the time. In addition, teeth had already begun to shrink from their peak size around 3 to 4 million years ago, when the only option for dinner might have been a raw tuber. Starting around 100,000 years ago, as humans began to develop better food-processing skills, much of the selective pressure for large molars was reduced, and selection started for smaller teeth that would fit better in a shrinking jaw.[21] By 10,000 years ago, teeth had shrunk even further, providing an even better match for the size of the human jaw at the time.

Hunter-gatherers still had a fair amount of chewing to do. This daily grinding of food was, perhaps, the most crucial factor leading to their well-aligned teeth. Because of all the chewing, the jaws of early humans developed more fully than a modern jaw reared on mac and cheese, applesauce, and smoothies. As with so many aspects of the human body, the development of the jaw plays out in a "use it or lose it" scenario.

In other words, if a human jaw is put through the evolutionarily appropriate paces, it may develop and function better than one reared on a diet of soft baby food and overly processed meals. Evidence in support of this idea comes from the work of Daniel Lieberman, a paleoanthropologist at Harvard University who has spent his career studying how evolution has shaped the human body. Lieberman and his colleagues tested this hypothesis about jaw development by raising two groups of hyraxes (little mammals that look like rodents but are actually more closely related to elephants) on either hard or soft diets.

The hyraxes on the mushy diet developed smaller jaws than those raised on the harder diets. In an article published in the *Journal of Human Evolution*, Lieberman and his team argue "that human faces may have become relatively smaller despite increases in body size because of reduced levels of strain generated by chewing softer, more processed food."[22] The switch to a soft, processed diet was the final straw that broke the back of the tooth–jaw marriage. Human teeth have continued to shrink in more recent millennia, but they have not been able to keep up with a small jaw that doubled down on its reduced size in the absence of any work to stimulate its development.

Of course, the real victim in this story was never *Homo sapiens* with our malocclusions and impacted wisdom teeth. The real victim was the mastodon. Very shortly after humans developed the tools and techniques that allowed for hunting and cooking, mastodons vanished from the face of the earth, sometime around 10,000 years ago.[23] Mastodons may have been wiped out by climate change, but it's also possible mastodon stew was the most delicious dish humans had ever made, and they quickly hunted the mastodons into extinction.

So is the modern tooth–jaw mismatch a result of our evolutionary past or is it due to our tendency to not work our jaws during the developmental years? Of course, those options are not mutually exclusive. Likely the answer is that both factors are at play. The modern jaw is both smaller than its predecessor and not put through the necessary chewing paces that would allow it to realize its full potential.

I would like to see an experiment that divided children by diet into soft and hard food groups and tracked their oral health into their teenage years. As I wait for the results of such a study, I'll hedge my bets and try to slip my young daughter some beef jerky and as many raw vegetables as I can get her to eat. Given her already solidly established love of oatmeal and yogurt, the more pragmatic approach is probably to start squirreling away money for braces as soon as possible.

2

The Fish-Eye Lens

After developing cataracts, which painter had one of their lenses removed, leaving them with the ability to perceive ultraviolet light?

a. Claude Monet
b. Vincent van Gogh
c. Frida Kahlo
d. Leonardo da Vinci

Childhood pictures of my wife, Julie, are adorable. Growing up in Montana, she always had a thick jacket on, and you can barely see her face for the giant 1980s Coke-bottle glasses she's rocking. Her parents figured out her vision might be a little off when she kept misidentifying sheep as cows. Keeping your cows and sheep straight is an important skill in Montana. At four years old, she was 20/200 in her right eye and 20/400 in her left. We met in college, and shortly after we started dating, her parents saved up the money and gave her the gift of Lasik eye surgery. She came out of it 20/20 in each eye. I know it was science, but it felt like magic. It's been more than 20 years since then, and she has yet to go back to glasses or contacts.

Julie's vision growing up was particularly bad, but she is certainly not unique in needing help to see clearly. A meta-analysis of more than 60,000 European adults revealed that more than half had some type of visual deficit.[1] Of course, as individuals get older, their likelihood of needing glasses increases significantly, which certainly inflates the data. It is a downright unique senior citizen able to reach the golden years without needing at least cheaters.

Visual difficulties are, however, not limited to the elderly. Approximately 25% of children in the United States use some type of vision correction.* The number goes up to roughly 40% of individuals for those in their 20s and 30s. Around the age of 40, the visual acuity numbers start to fall off the charts. Or, rather, the chart starts to look very fuzzy. After the age of 50 or 60, the only people left with uncorrected vision are genetic freaks like former professional baseball players who had 20/10 vision to begin with. But the point is, even at an early age, many individuals need correction. Right from the start many humans are set up with lousy vision. After millions upon millions of years of evolution, why are so many of us cursed with such a blurry view of the world?

---

*Reports from nonprofits such as the Vision Council and from federal groups like the National Center for Children's Vision and Eye Health indicate that approximately one in four children suffer from either myopia (nearsightedness) or hyperopia (farsightedness), with many of those children also having astigmatism.

BUILDING A CAR OUT OF A BOAT

To begin to get at an answer, imagine for a minute the life and career of a shipwright named Charlie. Charlie is no ordinary boat-builder. He possesses an incredible degree of technical skill and attention to detail. Charlie's superiors put him in charge of building a new, trendsetting boat intended to push the technological envelope. Once complete, the new craft will truly revolutionize the boating industry. It is quite the ambitious endeavor, and Charlie works on the project diligently for several years. Even the earliest versions of the new boat are fantastically different and exciting, but Charlie continues smoothing the wrinkles, adding some spiffy elements, and working every day toward making the boat better. One day his boss comes down from his upstairs office, and Charlie decides it is time to reveal his masterpiece.

Before he can rip the tarp off and display his life's work, his boss says, "So, I just got off the phone with corporate. Change of plans. Turns out they don't want to take this beauty out on the water anymore. There have been too many shark attacks lately and no one is buying boats. They want it to work on land. Oh, and you can't start over. We have way too much money sunk into this thing. I'm gonna need you to tweak what you've got there and turn it into a car."

Charlie lets out a long sigh, threatens to quit, and swears under his breath. But because he is a good employee, and because he has bills to pay, he eventually just shakes his head and gets back to work. Time passes and Charlie does his best to make a car out of what was supposed to be a boat. It looks kind of funny (like one of those duck boats used for shuttling tourists around in Boston or Seattle), and it is definitely not what he would have made had he scrapped the original project and started over. It does not work perfectly, but it works well enough, and Charlie retires satisfied, knowing he did the best he could given the ridiculous task of making a car out of a boat.

The short tale of Charlie and his career is a metaphor for the evolution of eyes in terrestrial vertebrates. Vertebrate eyes originally evolved to view objects underwater, and now we are stuck with them on land. Just as Charlie could not predict what his boss would demand, the process of evolution does not plan for the future. When vertebrates started living out of the water 375 million years ago, they already had eyes that had been around for 100 million years. They could not just scrap their old ocean eyes and start from scratch, working on new land eyes. The slate does not get wiped clean when animals take on a new environment. The early vertebrate land pioneers already had fully functioning eyes. They did not work terribly well out of the water, but they worked a lot better than nothing.

That, right there, is one of the most important features of evolution. Lousy function trumps no function. Every time. In the competitive coliseum of evolution by natural selection, a frog with blurry vision is going to win out over a blind frog 10 times out of 10. So life kept grinding along with what was already in place, and eventually, after countless generations of natural selection, the eyes became much better at doing their job on land. The result is that our eyes are now much better at viewing the world above the water than they are at trying to make out details underwater. After all, 375 million years is quite a long time to knock out a few kinks. In other words, even though it started off as a boat, our duck boat is now, in modern times, a vastly better car than it is a boat.

PRIMITIVE EYES

The eyes are basically extensions of the brain out at the very edge of the skull. The optic nerves connecting the eyes to the brain are, in fact, pieces of central nervous system tissue, like the brain itself. At the back of each eye is a retina, a patch of light-sensitive cells that generate a signal sent to the brain via the optic nerve. The retina has the most important job in the eye. Without a cornea or lens, an eye is only capable of achieving blurry vision. Without a retina, an eye is blind.

Studying the evolution of eyes is tricky. Eyes are soft and squishy and, as such, do not fossilize worth a lick. It takes a long, long time for something to turn into a fossil. This is no problem for structures like teeth or bones, which can sit around in the sun and rain for an awfully long time before they break down. Long before they can ever fossilize, most dead eyes are gobbled up by some opportunistic scavenger or they are simply turned into soup via the natural bacterial-driven processes of decay and decomposition. Decomp soup does not leave evolutionary biologists much to study.

The solution to this problem is to study extant evolutionary relics. Conveniently, there are living examples of very primitive animals wandering the earth to this day with their supremely old-school eyes. Back into the ocean we go to revisit the lovely hagfish. (Let's hope there are a few left that have not been made into wallets or belts.) Hagfish have fantastically ancient eyes. They consist of small sets of photoreceptive cells (basically very simple retinas) sitting beneath layers of skin thin enough to let light through. They totally suck as eyes. The Australian authors of a very well-organized and detailed research paper about the evolution of eyes in vertebrates address this point by bluntly stating, "Behaviourally, the hagfish seems to be almost blind, and its weak response to light is unaffected by removal of its eyes."[2] Clearly, their eyes are not very advanced if they can be removed without dramatically affecting the animal!

For a hagfish, what's the point of having eyes then? As with the origin of all anatomical features that eventually became complex, eyes started off as very basic, humble structures. The earliest eyes did not form an image. They simply detected the presence or absence of light. They were like the light-sensing gizmo I plug my Christmas lights into so they'll turn on at dusk. The sensor does not form an image; it's not like a security camera. Some animals use primitive light-sensing eyes to tell them when it's safe to crawl out from under a rock. Hagfish researchers think hagfish use their primitive eyes to set their internal clocks. Having biological rhythms (either daily, monthly, or annually) is an important part of animal life, and scientists think hagfish use their ability to sense light to help control their circadian patterns.

Complexity builds in the evolutionary cauldron in this manner. A structure starts off with one purpose and then, with the accumulation and selection of new mutations, it slowly becomes more sophisticated. As the structure changes, its purpose can change entirely. As the eyes became more complex, their function changed from setting a clock to forming an image.

For vision to be clear, the light needs to land directly on the retina within each eye. Myopia, or nearsightedness (when vision is clear up close but blurry far away), happens when the light focuses into a single point in front of the retina. Hyperopia, or farsightedness (when vision is clear far away but blurry up close), occurs when the light focuses into a single point behind the retina. It's unusual, but it is even possible for someone to be myopic in one eye and hyperopic in the other. Astigmatism is when the light does not focus into a single point, leading to blurry vision from all distances.

Moving forward to a still quite primitive fish like a lamprey, we pick up many more of the accessory visual features, like the cornea and lens, which focus the light onto the retina. With each step forward (or more appropriately, each swim forward, since all of this anatomy first evolved in the water), the eyes achieved better image resolution. Over time the eyes changed from simple light sensors into security cameras.

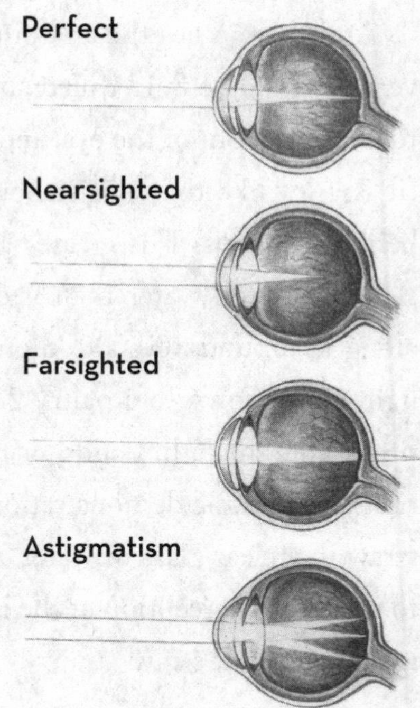

BENDING THE LIGHT

There is one particular feature of the eyes that is a clear reminder of the watery past of all vertebrates. This feature comes up each year when we dissect cows' eyes in my anatomy and physiology courses. Students have to work a bit with scalpels to cut through the outer layers of the eyes. It is worth the trouble because the most interesting anatomy is on the inside where the lens, iris, and retina are located. Even though I always stress the need to go slowly and make small, careful incisions, there is always at least one student each year who attacks the dissection with unbridled enthusiasm. Their reward for such vigor is a blast of fluid squirting up toward their own eyes. Nothing like a shower of bovine aqueous humor to stress the importance of lab goggles.

And therein lies the anatomical rub with the vertebrate eye. A very watery-like fluid called aqueous humor bathes the structures toward the front of the eye, and the more jelly-like vitreous humor (it is a lot like the consistency of egg whites) bathes everything behind the lens. This creates an issue because light travels more slowly through water than it does through air. Instead of racing along at around 300,000 kilometers per second as it does in air, light slows down to a paltry 225,000 km/sec in water. Upon hitting water, the light bends, or refracts, in the parlance of optometrists. The classic demonstration of this phenomenon is to place a straw in a clear glass of water. As the light refracts going from air to water, the perception at the intersection of the air and the water is one of a bent straw.

Another more dramatic way to demonstrate refraction is to run a small experiment. The materials needed for the experiment are a notecard, a marker, a clear empty glass, and some water. Not surprisingly, you'll find an example of the demonstration on YouTube if those materials are not readily available.* Draw an arrow on the notecard and lean it against something so it stands up. Place the empty glass in front of the arrow, leaving at least a foot between the glass and the notecard. Look at the arrow through the glass. Now, fill the empty glass with water and look through it again. The arrow changes direction! When the light's rays bend upon hitting the water (and the glass, twice) it creates an optical illusion, and we perceive the arrow as reversed.

In fact, even before adding the water, the arrow should have gotten fuzzy when looked at through the empty glass because glass also has the effect of bending the light. Light travels even more slowly through glass (somewhere around 200,000 km/sec) than it does through air and water. Many of us take advantage of this fact by placing glass in front of our eyes to compensate for the imperfect job our corneas and lenses do in bending the light. Of course, the glass needs to have the right shape to perform the right correction. Getting the exact right shape is not a trivial matter, which is why optometrists make a good living and why prescription glasses are a lot more expensive than, well, plain old glass.

---

*Search glass refraction arrow to find videos of the phenomenon on YouTube.

FROG GOGGLES

Why is there water in our eyes if it refracts the light, leading to potential complications? Why did dry eyes not evolve, eliminating the issue of the different refractive indices of air and water?

The answer, as you might expect after the introductory metaphor about duck boats, is that our eyes originally evolved in water. In the ocean, watery eyes solved refraction problems rather than causing them. The light bent upon entering the water, but the fluid within the eyes kept it from bending again. Having air-filled eyes would have meant two distortions of the light. Some of the fluid and nearby cells hardened up into a lens, which helped focus the light and provide an unforeseen degree of visual acuity. The underwater world became one of see or be eaten. The most effective predators had the best eyes and passed on their traits of great visual resolution and acuity. The potential prey with the best vision survived the onslaught because they could see the trouble coming and find somewhere to hide. They lived, mated, and passed their skills down to their offspring. There is nothing like a good anatomical arms race to speed evolution along.

Eyes were well on their way as effective structures when some ambitious fish tried their luck at spending time out of the water. The light that had been landing directly on each retina before was now all over the map. The refractive correction served by having wet eyes was now a refractive hindrance.

Exactly *why* some fish first got out of the water is a hotly debated topic. Saying they did so to avoid aggressive, toothy marine animals is likely a gross oversimplification. Some of the earliest research on the subject pointed toward early land vertebrates adapting to a life out of the water as a way of dealing with aquatic environments that ebbed and flowed.[3] A dried-up pond very quickly applies some serious selective pressure. Predator avoidance may also have been a part of the story, but others have argued that increased foraging opportunities within the intertidal zone or the shallow waters of a wooded floodplain drove the process.[4] Watching an extant fish like a mudskipper leave the water and suck down prey on beaches exposed during low tide makes for a compelling argument for the foraging opportunities hypothesis.*

Mudskippers seem perfectly happy as fish out of water.

The ultimate answer about the origin of land vertebrates likely lies in some combination of events. Regardless of why some fish left the water, what matters for the discussion here is simply that they *did*. The transition from water to land sealed the fate of our eyes.

---

*Search mudskipper on YouTube to see an array of videos of mudskippers foraging in the mud and jumping around on land to impress their potential mates.

It's been an uphill battle with our vision ever since that moment. Now, however, we have been on land for so long our vision in the water is terrible. To effectively see any detail in water requires us to simulate a *land* environment by surrounding our eyes with air, as we do with swim goggles and masks. The first amphibians probably would have been much better off if they had been able to wander around with water-filled goggles.

The fluid in our eyes has more of a job than simply bending the light. It is also critical for maintaining intraocular pressure. Without the pressure created by the fluid, an eye would collapse like a deflated balloon. If a cornea dries out it quickly becomes damaged, causing pain and irritation. A dry retina is useless. Take a lens out of its vitreous bath and it is no more effective at properly bending light than a marble. We have a suite of anatomy and behaviors in place to keep all the parts of our eyes from drying out. It has been a constant battle ever since our distant ancestors left the water behind.

THE SCIENCE OF BLINKING

First and foremost, the windshields of the eyes, the corneas, need to stay moist, and we spend a shocking amount of energy making sure they do not dry out. Blinking keeps the windshields of the eyes wet. If eyes do not blink even for just a minute or two, they quickly lose their ability to focus. We blink, on average, about 15 times a minute when awake. Running the math for a 16-hour waking day comes out to more than 14,000 blinks per day. When we are awake, we spend up to 10% of our time with our eyes temporarily shut, running the windshield wipers.

Blinking that much requires a lot of windshield wiper fluid. The lacrimal glands situated above the eyes produce the bulk of the fluid. Like windshield wiper fluid, tears are not 100% water. They are a cocktail solution. Different circumstances call for different blends of the fluid. Basal tears are filled with salts and other solutes and are in charge of keeping the corneas wet. Extra salt is also shed in tears when the salt concentration in the body gets too high. Some mucus is mixed in to help the basal tears coat the corneas. There are also some antibodies and antibacterial enzymes present to help fight off any infection trying to use the eyes as a point of entry. Any entry point into the body requires a heightened level of defense. The mostly aqueous solution produced by the lacrimal glands picks up some oils made by other, smaller glands around the eyelids. The oils are critical for trapping the tears next to the eyes. Without the oils the tears would run down the cheeks, which would defeat their purpose (and prove socially awkward).

All of those tears need to go somewhere. That's where the tear ducts come in. Like most of our anatomy, tear ducts are one of those structures we rarely think about until they stop working properly. Tear ducts are drains at the inner edges of the eyes. Under normal conditions tears flow through the drains and down into the nasal cavity. The nasal cavity connects to the oral cavity and throat (as adventurous children discover creatively with a long piece of spaghetti), and most of those tears end up getting swallowed. At a total of around 300 milliliters (~10 oz) per day, it is not an insignificant amount of fluid. That is almost a Coke can full of tears every day. A blocked tear duct can be a miserable experience, leaving the corner of the eye swollen and inflamed.

There are other times when the lacrimal glands go beyond their basic job description of keeping the eyes wet. The glands will crank out a different blend of tears called reflex tears when they detect the presence of an irritant. Those tears consist almost entirely of water and flow when a grain of sand gets in your eye or when you are chopping onions. Their job is one of flushing the eyes to clear out irritants and debris.

The third type of tears are emotional tears. They might come out during a tearjerker movie, or in the case of my daughter, they flowed this morning when I told her she had to wear socks and shoes if she wanted to go outside. (She strongly disagreed with the terms of the deal, and the waterworks started. In my defense, we live in Idaho and it is currently the middle of winter.) In such extreme examples of emotional distress, the lacrimal glands crank out *way* more fluid than is needed for simply wetting the surfaces, and the drains are unable to keep up. The emotional tears run down the cheeks, and if the event progresses to full-on sobbing (which, thankfully, it did not this morning), some tears even flow out of the nose instead of going down the throat (the nasal cavity also being connected to, not surprisingly, the nose).

Interestingly, women have smaller drains for tears than do men. This explains why their tears are more likely to overflow the metaphorical bathtub. Guys are anatomically set up to suck it up and reclaim all those tears. Emotional tears are also a different blend from those used for run-of-the mill eye wetting or for reflex crying. The emotional, sniffling variety contain hormones like adrenocorticotropic hormone (ACTH) and prolactin. ACTH is a stress hormone, and prolactin is most commonly associated with the ability of female mammals to produce milk, though it has a wide range of functions in both sexes. No one knows exactly what those hormones are doing in tears, but don't worry, the blink scientists are working on it.

Speaking of the blink scientists, they recently figured out that we blink a lot more frequently than is necessary to simply wet and clean our eyes.[5] What's with all the excessive blinking? When researchers started to look at blinking frequency and timing, they discovered something very interesting. People do not blink at random times. For example, when reading, people are more likely to blink when they reach the end of a sentence. When having a conversation, people subconsciously time their blinking to occur during a pause or a lapse. The thought is that blinking somehow helps the brain reset its focus. Blinking may allow us to draw a black curtain on the last idea or piece of a conversation and bring a fresh level of attention to whatever comes next.

WINDSHIELD WIPERS

A full reservoir of windshield wiper fluid doesn't do much good if the car doesn't have windshield wipers. The windshield wipers of the eyes are the eyelids. Eyelids and the fluid they draw over the eyes are necessary only because we no longer live in the water like our ancestors did. Most fish do not have eyelids because, living in water, they never need to blink to keep their eyes wet. In fact, they never close their eyes. Fish still manage to get some restful snoozing in each day; they simply sleep with their eyes open. We close our eyes when sleeping only to keep them from drying out.

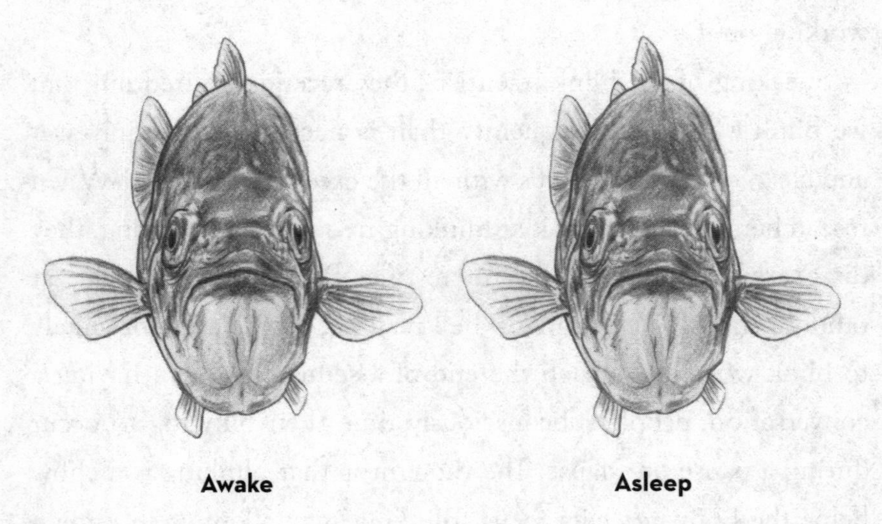

Eyelids first showed up in force in amphibians. Even if the first iterations of eyelids were rudimentary, amphibians would have been able to get away with it because they stayed near water. The earliest frogs and salamanders could have just hopped or crawled back into the water if they felt their eyes were drying out. In order for the bold reptiles to move life completely away from the water, they needed fully formed eyelids, or they had to employ other, more creative options for keeping their eyes moist. For example, most species of geckos don't have eyelids, so they are unable to blink. Instead, they resort to licking their eyes on a regular basis to keep them clean and wet. I'm not sure I like that option any better than blinking 14,000 times a day, but it is a rather innovative solution to the problem. Remember never to enter a staring contest with a gecko, should the occasion arise.

Any discussion of eyelids in animals must include nictitating membranes, which are translucent or transparent eyelids that can cover up the eyes for protection while still maintaining vision. Present in most terrestrial vertebrates (and some fish, like sharks), nictitating membranes cover the eyes while their owner is acquiring a meal. They protect the eyes of carnivores from thrashing prey and the eyes of herbivores against scratches from grasses and shrubs as they chew plants with their heads down. Primates lost their nictitating membranes for reasons not fully understood (likely driven by a change in diet). We are, however, clearly descended from animals with nictitating membranes because we have pink, nub-like structures that are remnants of the leftover third eyelids visible at the nasal corners of our eyes.

Most dog and cat owners have gotten a peek at this third eyelid, also called a haw. It is sometimes possible to see the haw when a furry companion is getting ready to settle in for a serious nap. A pet's haw may also be visible after a fight in which one or both of their eyes get beat up. A cat or dog will draw the haw over their eye after an injury to protect it during rehabilitation. Haws are also hugely beneficial for cats and dogs during a fight. They act like a pair of goggles, allowing animals to thrash away from close range without having to worry about getting their eyes scratched out.

Lastly, nictitating membranes do a fabulous job of keeping eyes moist. Keeping the eyes wet via the third eyelids means less time spent blinking when hunting. For an owl or a cat, blinking can mean the difference between successfully catching dinner and going hungry. Those mice are fast, shifty critters, after all. One blink and they're gone. Nictitating membranes are also particularly helpful for animals like camels, who need to cover their eyes because nearly every piece of the environment is small enough to sneak in and cause irritation.

NIGHT VISION

Lest we lose focus here, recall that the whole point of this chapter is for us to try to understand, from an evolutionary perspective, why our eyes do not work quite as well as we would hope. We've so far addressed the ways vertebrate eyes have changed over time to work in a dry environment after initially evolving in the ocean. The picture is not as sharp as it could be as a result, and we are stuck moistening our eyes nonstop anytime we want them open. Imperfect resolution and being a slave to staying moist are not, however, the only shortcomings of human eyes. As we turn our attention to human deficits in both night vision and color discrimination, this would be a good time to blink to reset your attention.

In addition to often having a blurry view of the modern world during the day, our visual abilities become downright pathetic once the sun goes down. Once light hits the retinas in the backs of our eyes, there are two different kinds of cells it can excite: rods and cones. The rods have a simpler job than the cones. They detect the presence of light (is there light, yes or no?) and relay a pattern to the brain interpreted in black and white. Rods are the cells eyes use under conditions of low light, and they are sensitive to all wavelengths of light across the portion of the electromagnetic spectrum considered visible light.

When bathed in excessive light, rod cells become washed out and unusable. When exposed to copious amounts of light, the photosensitive pigment made by the rods (rhodopsin) breaks down at a faster rate than it is put back together. Upon plunging into darkness, rods take some time to recharge, which explains why it takes a few minutes upon entering a dark room for our night vision to kick in. Upon recharging, rods take over entirely, and the cones take a rest until the lights come back on.

Nocturnal hunters have experienced strong selective pressure for increased rod density and, thus, ended up with eyes loaded with rod cells. This increase did not happen over the lifetime of one animal, but instead, over the course of many, many generations. For night dwellers, greater rod densities and better night vision would have translated into more successful hunting and consequently more food. More food would have translated into more surviving offspring. Having descended from individuals with great night vision, those surviving babies would have had retinas packed with rod cells. It is a tight little example of evolution by natural selection.

As the rod density increases, the cone density decreases (there's only so much room back there in the retina), but lousy color vision does not matter for nocturnal hunters because they are asleep under a bush for most of the day. Humans devote vastly more retinal real estate to cones compared with nocturnal animals. This has allowed us some amazing color vision during the day but has left us blind as a bat at night. Even that comparison is totally unfair to bats because they are decidedly not blind and can also echolocate as an added bonus. The expression should be "blind as a human at night."

There is another reason our night vision is lousy compared with so many other mammals. While performing the cow's eye dissection, if my students get past the stage with squirting fluids, one of the most fascinating aspects of the anatomy is the brilliant blue membrane coating the back of each eye. The color of the tissue is incredibly vivid; it looks like a little patch of Caribbean ocean sitting there behind the retina. The layer is known as the tapetum (= tapestry) lucidum (= bright). It does look somewhat like a pretty tapestry, but its function is closer to that of a mirror. Some of the light that reaches the retinas misses the photoreceptive pigments. In the absence of tapeta lucida, any light able to sneak past the retinas without stimulating the cells has no way of generating a signal. The tapetum layers reflect the leftover light back to the retinas, giving the light a second shot at stimulating the rods and cones. Humans don't have these layers, so we get only one shot for the light to generate a signal.

In animals with bright tapestries, some of the second-chance light manages to sneak past the retinas again, at which point the reflected light comes back out the fronts of the eyes and leads to the phenomenon of eye shine common in many mammals. Even when exposed to bright light, human eyes do not shine back like those of a possum, cat, or raccoon. Without tapeta lucida, in humans, bright light reflects off the many blood vessels at the backs of the eyes, leading to the red-eye effect commonly seen in flash photography. Cameras with anti-red features use short flashes of light, before the real flash, to get the irises to constrict and make the pupils small, limiting the amount of light from the real flash that gets to the backs of the eyes.

Like rod cell density, presence or absence of tapeta lucida tracks cleanly based on whether or not animals benefit from the increased sensitivity. Most fish have them because most of the available light is scattered and absorbed by the water, and thus, getting enough light onto the retinas is a constant battle. With land animals, it depends on whether they are snoozing or racing around at night. Animals that chase mice all night possess tapeta lucida and have killer night vision. Humans descended from animals that likely had the reflective layers, but they were lost in the primate lineage when primates went down an evolutionary branch that had them foraging during the day and sleeping at night.

TRICKED-OUT GENES

Since our vision is basically garbage at night, you might think we would make up for it with incredible color vision during the day using our relatively greater cone cell density within our retinas compared with many other animals. There is some truth to that, but in terms of color discrimination, we still cannot compete with some of the humblest vertebrates out there, such as fish and reptiles. To understand why, we have to look at the backstory of cone cells and how they cause different types of color vision in different animals.

Unlike the pigment produced by rods, which makes rod cells broadly sensitive to light, cone cells produce pigments sensitive over select portions of the electromagnetic spectrum. Humans and other primates typically have trichromatic vision, which means we have three different types of cone cells, each producing a single pigment with differing maximum levels of absorbance. The pigments are typically described as blue, green, and red, though in reality the maximum levels of absorbance are closer to colors we would call violet, green, and green (the second green being tinted with a fair amount of yellow). We perceive all the varied colors by the firing of a combination of those three types of cells. Most other mammals have only two different pigments, and as such, we like to think of our group as being the most highly evolved in terms of color vision. A little bit of digging into other animal groups, however, shows even our color vision has some serious limitations.

This is one of those funny cases where you have to look back down the tree of life to find increased complexity. Rods and cones have been around a long time. They crop up very early in vertebrate evolution, already present by the time the super ancient hagfish and lampreys were cruising the ocean 450 million years ago. The rod cells put their proverbial heads in the sand and stuck with the one type of photopigment. To this day their job is still simple—is there light, yes or no? The evolution of cone cells is where light detection gets interesting. The cone pigments produced by cone cells broadened out into several different varieties through a series of gene duplication events.

Gene duplication is a lot like it sounds. Roll back the clock far enough and there was only one type of cone cell producing one type of cone photopigment. A gene in the DNA of some ancient fish coded for production of that one photopigment. At some point a *copy* of that gene became established in the genome of a primitive fish, giving them two copies of the gene controlling the production of the cone photopigment.

Copies of genes regularly end up scattered throughout any given genome because DNA is replicated in cells all the time; it is a fundamental step in the process by which cells divide. Even the simplest organisms are composed of cells with millions of base pairs (the building blocks of DNA) in each cell. Human cells have three billion base pairs in each cell. *Billion*, with a *B*! Sometimes, during the process of replicating all that DNA, an extra copy of a gene or two slips in there. If those mistakes happen in the formation of a sex cell or early in embryonic development, those extra copies of DNA end up fixed into all or most of the cells of the body.

When copies of genes end up in the genome, it can spur evolution in interesting directions. To understand how, imagine a one-car family with a teenager who desperately wants to drive. If they own just the one family sedan, the car will probably stay the way the parents want—no funny bumper stickers, nothing hanging from the rearview mirror, and certainly no off-market speakers to kick up the music. Now, consider if one morning this family wakes up and there is a carbon copy of their boring sedan sitting in the driveway. They can keep the original all staid and adult-like, and the teenager can trick out the other one however they want. It does not take long before the two versions of the car become notably distinct. Mutations are simply more likely to stick when there is another version of the car (or gene) cruising along with its traditional function. Parents are much less likely to object to hot-pink seat covers when they don't have to sit on them when they drive around town.

In a rather short amount of time, evolutionarily speaking, starting with one type of ancestral cone photopigment gene, gene duplication led to *four* different classes of cone cells in some early fish.[6] Even before a few fish up and crawled out of the water, many of them had developed tetrachromatic (four-pigment) color vision. It is especially beneficial, and seen frequently, in shallow-water fish where being able to distinguish different color shades pays great dividends in terms of predator avoidance and feeding. It is clearly less beneficial deeper in the ocean (the light cannot get much past the surface), and as such, the deeper-ocean dwellers tend to have a simpler setup in terms of their cones. But once out of the water, the advantages of advanced color discrimination take off. Without the water sucking up the light, there is every color imaginable (and some that are unimaginable) waiting to stimulate eyes on land.

Many fish, reptiles, and birds still use the four-pigment color detection system. They see by using four classes of photopigments, compared with the trichromatic system of humans and the dichromatic setup of most nonprimate mammals. Each cone allows for the perception of around 100 different shades of color. So one cone allows for discrimination of 100 shades, two cones 10,000 shades (100^2), three cones 1,000,000 shades (100^3), and four cones 100,000,000 shades (100^4).

FIFTY (THOUSAND) SHADES OF GREEN

Why did most humans end up with only three types of cones while the lucky reptiles and birds are blessed with four? Also, why did most of the other mammals get stuck with only two, making their color vision even worse than ours?

The conventional wisdom has always blamed the dinosaurs. The logic goes that the best way for early mammals to avoid being bite-sized snacks for dinosaurs was to hide out during the day and be active at night. All those years spent snoozing during the day led to the loss of incredible color vision in early mammals. Recent evidence, however, suggests the story might be more complicated than this traditional narrative.

A team of researchers from the Field Museum and the Claremont Colleges discovered evidence of nocturnal activity in Paleozoic mammalian ancestors (called synapsids), predating the evolution of mammals by over 100 million years and the evolution of dinosaurs by more than 50 million years.[7] In other words, many mammal-trending animals were becoming nocturnal even before they were fully mammals and before there were dinosaurs to worry about. Diet appears to have driven the change. Some synapsids were moving to a herbivorous diet, and the timing of the switch coincides with the shift to a nocturnal life. Maybe there was less competition running around at night looking for food. Regardless of exactly when or why, at some point mammals went for the up-all-night, sleep-all-day, rock-and-roll lifestyle.

For animals that remained diurnal during this whole period, any negative mutations in cone photopigment genes were weeded out by natural selection. There would have been strong selection against even the slightest decrease in visual discrimination. In mammals, those negative mutations could stick because when they popped up in the genome, those genes were inactive. The mammals were working the evolutionary graveyard shift, and it did not matter at the time if there were mistakes in the cone photopigment genes.

Once dinosaurs evolved, there may have been a little extra motivation for small proto-mammals to scurry under a bush and hide through the day. After most of the dinosaurs went extinct, many mammals went back to a diurnal life. Then, it would have been nice to use all four photopigment genes, but two of the genes had accumulated so many mistakes while mammals were nocturnal that they were no longer functional. Mammals with tetrachromatic reptilian ancestors had come out the other end of the evolutionary tunnel as dichromatic.

The primate lineage was able to recover, to a degree. One of the two remaining photopigment genes duplicated again in primates, leading to the trichromatic vision seen in humans and other primates to this day. There has been a fair amount of debate as to why trichromacy was able to establish itself in primates but has remained out of the grasp of the bulk of other mammals. Again, diet seems to have been the driving factor. As leaf and fruit eaters, the ability to distinguish between different shades of green and different degrees of ripeness in fruit would have been hugely advantageous for early primates. Even though we have only three color pigment types, the most recent split set us up with a keen ability to discriminate more shades of green than any other color. We can distinguish tens of thousands (some even argue hundreds of thousands) of shades of green, and we have those early salad-eating primates to thank for this now-much-less-necessary skill. These days it's maddening because it means there are way too many color swatches available when deciding which exact green to paint the guest bedroom.

BEYOND THE RAINBOW

Not every human is a trichromatic primate. Color-blind individuals produce two fully functional types of cones and one mutated version that is deficient in its ability to discriminate either red or green. This does not mean color-blind people see in black and white. It just means they cannot distinguish as many shades as someone who has three fully functioning cone types. They are dichromatic, walking around seeing a limited number of shades, like they are looking through the eyes of a dog or a grizzly bear.

Color blindness is much more common in men because the genes controlling the production of the color pigments are on the X chromosome. Men have only one X chromosome. The other sex chromosome in men is the Y chromosome, which does not contribute to color vision in any way whatsoever. With only one X chromosome, if males get a color-blind X, they don't have another X chromosome to fall back on, as women do. I have had dozens of male students who were color blind and, to date, only one color-blind female student who got a deficient copy on both of her X chromosomes. Color blindness is particularly common in men with northern European ancestry, affecting nearly 10% of males with that particular background. At my first teaching job at a very small college in northern Wisconsin (where many people have northern European roots), many of the male faculty members were color blind. I was the unusual one able to see the entire color spectrum in all its glory.

The color blindness story takes an interesting turn when looking at the daughters of color-blind men. Such women actually possess four different cone types. For the purposes of the example, let's say the mutated cone in this case is the green cone. A woman with a color-blind father would have two copies of the red cone (one from each parent), two copies of the blue cone (again, one from each parent), one copy of the normal green cone (from her mother), and one copy of the mutated green cone (from her father). That makes for four different cone types: normal red, normal blue, normal green, mutated green. Such a woman is theoretically tetrachromatic, just like a bald eagle or an iguana.

The idea that tetrachromatic women with extraordinary color perception might exist was first suggested in the 1940s but not given proper experimental scrutiny until more recently. Researchers from Newcastle University generated images that would look the same to trichromatic viewers but different to tetrachromatic individuals.[8] They studied 24 women who were the daughters of color-blind men and put them through a series of tests. One after another, the women were trichromatic, unable to make any distinctions beyond those expected by humans with normal, run-of-the-mill color vision. One woman, however, listed in the study as subject cDa29, was able to pass every single tetrachromatic test the researchers could throw at her. They had found the world's first known functionally tetrachromatic human. The scientists don't know, at this point, why her four pigment types have all remained active, in contrast with the other 23 women for whom the mutated cone is inactive, as it is in color-blind men. Also, no one knows how many other cDa29s are out there, but there are likely many given there are millions of women descended from color-blind men. Even if only one out of every couple of dozen is functionally tetrachromatic, that still would make for many, many women with extraordinary color vision.

Subject cDa29, reptiles, and birds with their four functional color photopigments are able to see shades of color the rest of us can't even imagine. I would love to talk to cDa29 and have her attempt to explain what the world looks like to her. I suspect it would not make much sense to a lowly trichromatic individual like me. After all, what words do you use to describe shades of color to someone who is unable to see those shades? All of the points of reference are other shades of color they cannot see. It would be like trying to describe the beauty of a rainbow to a dog. I've had that type of conversation with some of my color-blind students, and trust me, you get nowhere fast.

SEEING THE INVISIBLE

For me, the most interesting aspects of human color vision explore the colors we are unable to see. Our anatomy limits us to a rather tight range, whereas many other types of animals are able to see a broader section of the electromagnetic spectrum. This ability has been recognized for a long time in honeybees. The ultimate honeybee whisperer was the Austrian ethologist Karl von Frisch, who, in the early part of the 20th century, demonstrated the perception of UV light by honeybees through a series of quite elegant experiments. As it turns out, some flowers (e.g., black-eyed Susans) feature patterns of UV reflectance visible to honeybees that are totally invisible to humans. The patterns serve as nectar guides and act like bull's-eyes to help bees target in on the appropriate sections of the flowers. The flowers benefit from this because bees come away with a dusting of pollen that they then transfer to the next flower.

I should say those patterns are totally invisible to *most* humans. I heard a story during my graduate training from my thesis advisor about UV vision in humans. My graduate advisor was a brilliant naturalist named Tom Eisner, who had a long and distinguished career at Cornell University as a chemical ecologist.* Tom was also quite an accomplished photographer. He published much of the original research showing UV patterns in flowers.[9, 10] At some point, long before we met, his father had undergone cataract surgery. Following the surgery, with his new synthetic lenses, to his great surprise, Tom's father could see UV patterns on flowers!

This anecdote demonstrates that our inability to see UV has less to do with our photopigments and more to do with the filtering of light by our lenses. Our violet photopigment (the one typically referred to as blue, with a peak absorbance around 420 nanometers) *is* sensitive to UV light and can lead to the generation of a UV signal sent back to the brain for interpretation. Human lenses, however, typically filter out the UV light and prevent it from ever reaching the cone cells within our retinas. We think we are blind to UV because our bodies never give us the chance to perceive it. When Tom's father had cataract surgery in the 1970s, not every synthetic lens used for cataracts filtered out UV. Without his natural filters in place, he was able to perceive patterns of UV reflectance off some flowers.

The French impressionist master Monet had a similar situation.

---

*One of my earliest research projects was a study in which Tom Eisner and I collaborated with a group of chemists to investigate the compounds responsible for UV patterns on flowers. Tom photographed flowers in visible and UV light and transduced the UV into a blue shade to give an idea of what the UV patterns might look like to an insect. For a honeybee, those UV patterns blend with the visible light patterns to make colors that Tom referred to with names like "bee purple" or "bee violet" (because a honeybee doesn't see *just* UV reflectance, they see both UV *and* visible light reflectance).

Struggling with cataracts late in life, at the age of 82 he had one of his lenses removed. His surgery happened in the early 1920s, nearly 30 years before the first implantation of an artificial lens in 1949. Monet came out of the surgery, like Tom's father, able to see UV patterns out of his one lens-less eye. For the last few years of his life (he died in 1926), he painted with a perspective very few other artists had likely ever experienced. Art historians debate whether or not it influenced his art in any significant way. If his missing lens did affect his art, it seems the postsurgery works could be truly appreciated only by honeybees trained as art historians, given the blindness to UV reflectance of the general human population.

Or, given recent discoveries, maybe art-loving dogs could also judge Monet's late-life masterpieces. Those fancy honeybees and the tetrachromatic birds and reptiles have long been known to see UV patterns. Recent evidence suggests many mammals may also be UV sensitive.[11] Nearly 60% of UV light is able to pass through the lens of a house cat. The figure is greater than 60% in a dog. The numbers drop for cattle and deer and totally fall off for primates. Most primates transmit less than 10% of UV back to their retinas, and many, like us, block it out entirely.

Why did primate lenses evolve to filter out the UV? (*I* want to see those patterns on flowers!) My gut reaction, as a biologist, is to explain it from a DNA-damage perspective. We know that too much UV radiation is very damaging. We suffer from many different types of skin cancers due to UV exposure. It makes sense that the human lens evolved to protect the retina by filtering out the potentially damaging UV radiation. That story was easy to swallow until this recent discovery of the widespread UV sensitivity of mammals.

While filtering out the UV may provide a certain degree of protection,[12] another hypothesis (not mutually exclusive from protection) suggests human lenses work like a pair of ski goggles. A good pair of ski goggles filters out just the right amount of light, leaving the skier with much better contrast and resolution compared with skiing without goggles. Mutations producing lens-filtering molecules within the lens have stuck around in the primate lineage ever since we became diurnal. By filtering out UV, this new hypothesis suggests, we are able to achieve higher resolution for both distance and up-close vision.

Being able to see fine detail helps us discern subtle differences in plants. Such discrimination might make the difference between eating something that leads to a stomachache (or worse) and having a full, satisfying meal. We can also pick out details about a prey item even if it is hundreds of feet away. Now, in addition to continuing to employ those more traditional benefits, we use our unique skill of high-def image resolution in very different ways. For example, our incredible acuity translates into driving ability because it allows us to see trouble far ahead on the highway and then also easily check our speed or locate the correct button to change the radio station. The uniquely human (or at least primate) ability to focus on close images allows us to hold a book up close and pick out fine details. Most other mammals, if they learned to read, would need to wear cheaters.

THE MYOPIA EPIDEMIC

Speaking of holding books up close, there is one last topic to dig into in this exploration of the origins of the difficulties and shortcomings associated with human vision. Epidemiologists have noticed a disturbing trend in regard to our visual abilities. In short, our vision is getting worse. Rapidly. The rates of myopia in Western civilizations have doubled from those a couple generations back. For reasons not fully understood, eyes are becoming elongated during juvenile development, which causes the light to come to a point in front of each retina. This has always been a problem for some individuals, but it is increasingly becoming a problem for seemingly everyone. There are parts of East Asia where greater than 90% of young adults suffer from nearsightedness. As noted in a recent article titled "The Myopia Boom" in the esteemed journal *Nature*, some of the cases are correctable with contacts and glasses, but in other circumstances the problem is so severe it can cause "retinal detachment, cataracts, glaucoma, and even blindness."[13] What has happened recently to set so many people down this nearsighted path?

The classic explanation for myopia sufferers goes back hundreds of years to scholars complaining about their vision going south owing to their excessive time spent huddled over books. Such tales of woe stuck hard in the collective scientific consciousness and, despite being basically untested, survived as the most oft-repeated explanation to this day. We have added a new, unsubstantiated scapegoat to the lineup with the proliferation of personal screen time in the 21st century. There probably is not a parent alive who has not tried some version of the "give your eyes a rest" ploy to get their kids to look up from a cell phone or some other screen. Many people decided, without hard evidence, that all those hours spent squinting at a screen (or an old-school paper book) cause eye strain and impair ocular development.

This hypothesis, born of needling grandmothers and worrywart helicopter parents, has not held up upon closer inspection. Multiple studies have failed to draw a connection between book time or screen time and myopia risk. Multiple independent studies have, however, uncovered a different, statistically significant risk factor for myopia: time spent outdoors.[14, 15] Or rather, lack of time spent outdoors. Children who spend greater chunks of their day outside have a lesser risk of developing myopia than children who spend their days inside. It doesn't matter *how* they spend their time outside. The outdoorsy kids in the studies spent as much total time on screens as the indoorsy kids. They didn't have to be kicking a soccer ball or climbing a tree. Even if they were playing around on their phones, as long as they were doing it outside, they were less likely to become myopic.

It is a shocking result given the total buy-in to the eye-strain hypothesis. It makes a fair amount of sense, however, when viewed through an evolutionary lens (sorry, I couldn't resist). It has only been in the most recent of times that we have confined ourselves so significantly to spaces devoid of natural light. Of course our eyes would need great amounts of natural light to develop normally; it was input from natural light that guided the healthy development of the eye for hundreds of millions of years. It should come as no surprise that eyes do not develop well when exposed only to much dimmer artificial lighting. It would be like expecting muscles to develop normally in the absence of gravity or our sense of hearing to develop with only limited exposure to sound during the early years of life.

Where does all this leave us with our eyes? There is obviously not a thing we can do about the fact that vertebrate eyes originally evolved in water and now we are fully committed to this whole land thing. Short of licking our eyes like geckos, we are stuck blinking all day to keep our peepers moist. Our night vision will continue to be lousy even if we all transition to a diet of mostly carrots. We are set up to discriminate the arugula from the escarole, not to chase mice around after dark. No luck bringing back UV sensitivity either, short of going the Monet path and taking out a lens without replacing it. No, we are basically stuck with a life of looking for our glasses and gradually increasing the size of the font on our phones and tablets. One thing we *can* do is kick our kids off the couch and make them go outside. The fresh air will do them some good, and who knows, it might even decrease the chances they will someday need glasses.

3

Down the Hatch

Which food causes the greatest number of choking deaths in children?

a. grapes
b. popcorn
c. nuts
d. hot dogs

Admittedly, there are a few genetic freaks out there with all-natural perfect teeth and 20/20 vision. They did not suffer through braces during middle school and never got to experience the joy of having wisdom teeth pulled. They have never had to clean contact lenses or wonder, for the millionth time, where they left their glasses. But even people with straight teeth and flawless eyes have a few clunky anatomical faults. For example, everyone, at some point, has been enjoying a meal and had a piece of food or a little liquid go down the wrong pipe. This quirk is a universal shortcoming of human anatomy, and it seems completely backwards from an evolutionary perspective. A blocked airway is not a good situation. Even a brief interruption in oxygen supply to the brain can have life-altering consequences. As such, the body freaks out a bit and some coughing and gagging take place until the potato chip or grape comes back up and finds the intended tunnel.

If some sputtering doesn't work, the situation can quickly become serious. My dad was at a company picnic years ago when a woman at a nearby table started choking on a piece of meat. In serious cases of choking, the blockage is so complete that the victim cannot speak or cough. His colleague, Esther, had been enjoying a bite of barbeque and the next second her life was in danger. Lucky for her, Dad had recently taken a first aid class, and he sprung up and executed the Heimlich maneuver. The lodged piece of meat came shooting out and Esther lived to tell the tale. But why does this happen in the first place? Why are our lives in danger every time we sit down for a meal? Why, for the love of God, is the air tube so ridiculously close to the food tube?

THE DANGEROUS ACT OF EATING

The anatomical issues of the first two chapters might cause some discomfort, or require forking over a good chunk of change for braces and glasses, but this eating problem can be straight-up deadly.* Choking is a leading cause of accidental death across all ages.[1]

---

*Choking (or unintentional suffocation) comes in fourth in the US after the three leading causes of unintentional deaths: (1) poisoning (overdoses being the most common type), (2) motor vehicle accidents, and (3) falling. Google CDC 10 leading causes of unintentional injury to see the data.

The issue is of particular gravity for babies and toddlers, with their dangerously narrow airways.

Data on the width of airways at different developmental stages illustrate the problem. For babies between birth and two years of age, the average anterior–posterior tracheal diameter is just over 0.5 centimeters.[2] That is the diameter of a run-of-the-mill pea. A pea! The infant airway is so small that even a large pea cannot safely make passage. By the ages of four to six the issue has started to improve, but kids in this range are by no means totally out of the woods, with an average tracheal diameter of only 0.8 centimeters. By the time kids reach 12 to 14 years old, the tracheal diameter has jumped up to 1.3 centimeters, which is a hair smaller than the diameter of a AA battery (1.4 cm).

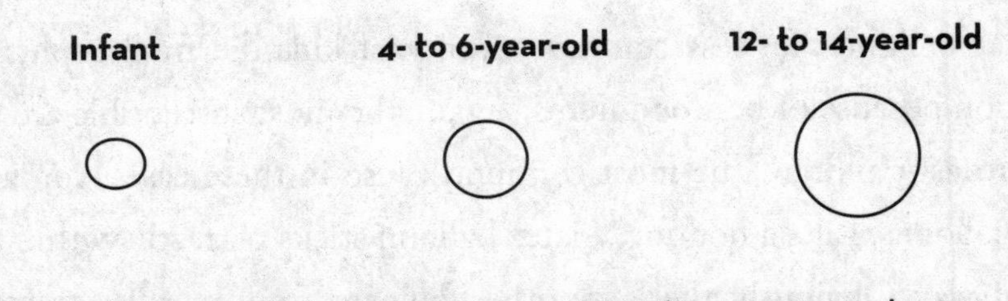

Actual tracheal dimensions by age. Slice those grapes and cut those hot dogs!

The problems do not stop there with small children and choking. Once something becomes lodged in the airway, the natural reflex is to cough. Little kids are unable to cough with the same force as adults. As such, it is difficult for them to dislodge foreign items. If the airway becomes completely blocked, death can result in a matter of minutes. The most common cause of a blockage is some type of food. Hot dogs make a tight seal on the airway and are the most common cause of choking in kids in the United States.[3] Nuts, hard candy, and grapes are other common culprits. Kids are also lousy at chewing (the muscles and coordination have not fully developed at an early age even if the teeth are present), and hot dogs create the perfect storm for choking—one that evolution had no shot at preparing us for. The solution, of course, because kids do love hot dogs, is to start by cutting hot dogs lengthwise and then cutting them again, and again, and again, until the pieces are so small there is no chance they could get stuck. It might seem crazy to hack a hot dog to death for a five-year-old, but given the airflow dynamics of their airway, it is the only safe option.

There is choking-related danger for kids beyond the dinner table. Albeit far less common, plenty of kids die from asphyxiation because of nonfood items. Again, the most susceptible are the littlest children. The most common cause in these cases, you ask? Balloons. Like a hot dog, a latex balloon sticks perfectly within the airway. Obviously, there are other dangers, from marbles to coins and everything in between. Some of the kids at highest risk are those with older siblings. The older kids leave toys lying around that are safe for the older children but not necessarily for the younger kids. Many safety advocacy groups have tried to simplify things for parents by recommending that small children should not have access to any toy capable of fitting inside a toilet paper roll.

The issue does not totally stop upon hitting puberty. The larger adult airway can still get clogged up, leading to hospitalization or death. The average tracheal diameter in a 20-year-old adult male is only 1.75 centimeters, which is still smaller than the diameter of most commercially available grapes.

As with so many biological features of life, the problem comes full circle, with the risk of choking rising again among the elderly. With the onset of events like strokes or conditions like Parkinson's disease, the neurological hookup between the brain and the tissues involved in swallowing can become clunky. A study looking at the risk factors involved in near-fatal choking episodes found those most at risk were the elderly and "those with neurogenic dysphagia."[4] Dysphagia means difficulty swallowing. The authors published the results in the journal *Dysphagia*. (Yes, there is an entire peer-reviewed journal dedicated to swallowing difficulties!) It is not exactly earth-shattering news that individuals who have difficulty swallowing are more likely to choke, but the study shines a spotlight on dysphagia as a profoundly serious issue for the elderly.

The threat of choking hangs over humans for their entire lives. Humans are most at risk early and late in life, but there is never a time at which there is not at least some risk to a person sitting down for a meal. Everyone should become well versed in the Heimlich maneuver, given the likelihood of choking. Since its development in 1974, the maneuver has saved more than 100,000 lives, including that of my dad's colleague Esther. Dr. Henry Heimlich had never personally used the technique to save a life until 2016, when, at the ripe old age of 96, he noticed someone in his seniors' home choking and came to their rescue, perfectly executing the maneuver that carries his name.

EVOLUTION OF THE LUNG

So what's up with all the choking? The anatomical intimacy of the trachea and the esophagus is the underlying cause of the problem. The trachea is easy to find when dissecting a mammal because it is white and contains unmistakable cartilaginous rings that help maintain the rigidity of the airway. Upon dissecting a rat, a fetal pig, or even a cadaver, my students often have trouble locating the esophagus, however, because it lies directly underneath the trachea, sticking to its posterior surface. The two tissues could not be situated any closer unless one of them was literally inside the other.

To understand how the trachea and esophagus became so wed to one another, we need to go back to the evolution of the lung itself. By 400 million years ago, fish had radiated and colonized freshwater habitats in addition to the ocean where they had initially evolved. A new kind of danger arose with freshwater environments. Namely, they can dry up, which is not exactly a concern in the ocean. A significant drought or even seasonal variation can lead to a pond or river becoming nothing more than a puddle of mud. For fish, the most common outcome of a dried-up water supply, both then and now, is a quick and floppy death. With gills exposed to air, fish that are able to extract oxygen only via their gills are quickly out of luck.

There were, however, a few species, or initially a few individuals within a few species, with unique mutations and characteristics that became strongly selected for when a pond or river dried up. Animals able to use their fins to move around on mud may have been able to get themselves back into a nearby supply of water. Such a strategy would have worked only if the next water supply was close, but it still would have been a considerable advantage over flopping around for a while and then dying.

An early fish able to gulp some air had the best trick up its sleeve. After all, there was not always another nearby source of water to stumble into. Fish able to survive out of the water did not give up their ability to use their gills; they were able to supplement their gill respiration with lung respiration when necessary. Lungs emerged in very limited groups of fishes from outpocketing of the digestive system. They were like little bedrooms coming off a hall leading down to the living room, or stomach. Some fish had a single bedroom coming off the hall, and others had multiple bedrooms.

Confusion arises in understanding the developmental history of the lungs because other structures called swim bladders or gas bladders *also* arose as outpockets from the digestive tract. By increasing or decreasing the amount of gas in a swim bladder, a fish with such a structure can control its buoyancy. (The swim bladder holds a dear place in my heart because it was the focus of my first dissection. I was around 10 when the family goldfish went belly up. I turned the tragedy into an elementary-school science project by dissecting out Goldie's swim bladder on the kitchen counter. I can still remember being surprised upon finding what looked like a perfect little balloon during my goldfish necropsy. Good on my folks for letting me at the family's dead goldfish for the sake of science.)

One older line of thinking suggested the two structures, lungs and gas bladders, evolved independently. This idea was put to rest with some elegant images of gas bladders and lungs in a variety of fishes, which indicate the two structures have a high degree of homology, or shared ancestry.[5] In other words, swim bladders and lungs are variants of the same basic structure. The earliest adopters of this new technology used their air-filled organs as lungs and fanned out in different directions, leading to many species of highly variable fishes. One group bailed on using the air-filled organ as a lung and it became a swim bladder. That group, the Actinopterygii, or ray-finned fishes, was wildly successful and makes up the bulk of fish species today (e.g., trout, bass, halibut).

The other group, the Sarcopterygii, played out the lung card and became lobe-finned fishes, including extinct fish like *Tiktaalik* and living species like the coelacanth and lungfish. Humans thought coelacanths were extinct for the bulk of our history. They went undetected by humans until the 20th century, hiding out in the depths of the ocean.[6] It was to everyone's surprise in 1938 when a South African fisherman pulled one up from the Indian Ocean. Living so deep, coelacanths no longer have the need for lungs, and as such, their lungs have become vestigial structures.[7] Lungfish stay around the surface and still very much use their ability to breathe air to get them through times when water is scarce.

The small group of lobe-finned fishes has not been nearly as successful as the ray-finned ones, but without this crafty evolutionary offshoot, none of the land vertebrates (including us) would be around today. Some particularly bold lobe-finned fishes eventually crawled out of the water, starting vertebrates on the path that eventually led to amphibians, reptiles, and mammals. This very odd group of fishes from hundreds of millions of years ago is responsible for many of the quirks of our respiratory anatomy.

BREATHING THROUGH THE NOSE

To answer the earlier question, the trachea and esophagus are so close to one another because the tube for one system (air) was born out of the tube for the other (food). The origin of the lungs via the digestive system is at the root of the choking problem. The issue is that the pathway leading to the lungs must cross the pathway leading to the esophagus. To understand why the paths cross, we need only look at the anatomical arrangement of the nasal and oral cavities. Not all lobe-finned fishes gulped air exclusively through their mouths. Some also breathed through their nostrils. This arrangement worked out very well for the amphibians who came later because they still lived near or in water. Amphibians must reproduce in water and are unable to wander far away from moisture, as they lack the tough, thick skin of reptiles. By breathing through their noses, early amphibians were able to keep their mouths underwater (where presumably most of their food would still have been) and use their nostrils to bring in air.

Humans have not lost this ability. Mammals have taken the separation one step further with the development of a secondary palate separating the nasal and oral cavities. If a frog has its mouth underwater and wants to breathe through its nose, it has to keep its mouth shut. With a secondary palate, a mammal can fill up its mouth with water and continue to breathe through its nose. Get a cup of water and try it out. All this nose breathing has the anatomical effect of causing the respiratory tract to start out *posterior* to, or behind, the digestive tract. This posterior positioning would have been well and good if the lungs were also toward the back. But they're not. The lungs are toward the front. Therefore, the airway must cross the digestive tract to get back to the front and eventually head to the lungs.

The intersection of the pathways for air and food is at the root of the choking problem in humans.

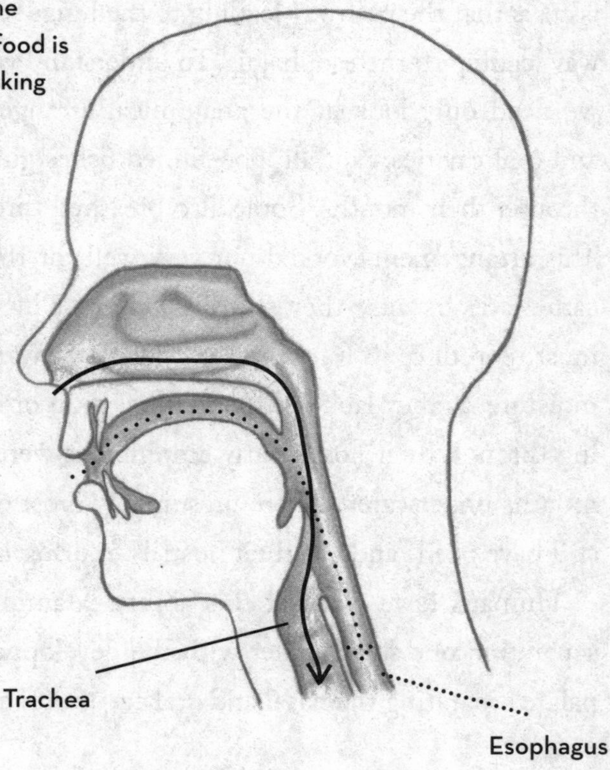

The problem would not exist had tetrapods (four-limbed vertebrates) evolved from animals with their air-filled organs positioned dorsally. The air could have come into the posterior and stayed in the back as it headed down to the lungs. Fish like carp and perch have a swim bladder oriented in this manner, and had frogs evolved from carp, choking wouldn't be an issue. But alas, land vertebrates' ancestral group were the lobe-finned fishes, with their anteriorly positioned lungs.[8]

THE DANGEROUS INTERSECTION

The intersection is the issue. Imagine you are driving down the highway. If you watch your speed and maintain a reasonable distance from other cars, it is a relatively safe way to travel. Now picture the road with a set of train tracks running alongside it. Again, there should not be a problem. No horrific pileups will occur as long as the train stays on its side and the cars keep it between the guardrails. Now pretend that, without warning, the highway and the train tracks cross. An arrangement of this nature does not always lead to an accident. The trains and the vehicles could sometimes, by dumb luck, arrive at different times and everyone would go on their way without incident. But when they happen to get there at the same time, the results are devastating. Train cars and minivans colliding at 100 kph (62 mph) does not make for a pretty scene. Similarly, most of the time when a little kid eats a hot dog it ends up sliding into the esophagus and heading down to the stomach. But sometimes the hot dog takes a wrong turn at the intersection, slips into the windpipe instead, and then our anatomy gives us a frightful reminder of our evolutionary history.

To continue the metaphor, there are times when roads and trains have to intersect. And I'm sure it did not take long before someone realized it would be a good idea to control those intersections. Simply putting up a couple of signs was a good start. With the addition of a gate and some lights, railroad crossings became much more organized. The same thing took place with the intersection of the digestive and respiratory anatomy. The traffic control features of the aerodigestive pathway are the components of the larynx, also known as the voice box. The larynx is the transition between the upper airway and the lower airway. A rudimentary larynx was already present in lobe-finned fishes simply as a sphincter closing off the bag-like lung.[9]

The sphincter that allowed closure of primitive lungs had developed from a gill arch in fish.[10] Also commonly referred to as pharyngeal arches, these loop-shaped bony features in fish provide structural support to the gills. Adult amphibians do not have gills and, thus, do not need gill arches to support gills. Instead of ending up on the evolutionary scrap heap, gill arches took on new jobs in land animals. Many of them became features of the head and neck in the tetrapods. The tiny bones of the inner ear are nice examples of structures with humble gill arch origins.

In addition to its more rigid features, the larynx also contains softer cartilages and membranous tissues capable of expansion and contraction. Some of those soft tissues can come together and create a seal within the airway. For an intersection having to now deal with air in addition to food, this was a critical adaptation. When not swallowing, the slit (called the glottis) stays open, and air passes freely down into the trachea and on to the lungs. When swallowing, the slit closes, which helps keep anything non-air from accidentally getting into the airway. The system is obviously not perfect, as sometimes a peanut or a grape can sneak past the glottis. This is no different from an accident occurring at a controlled intersection. Every now and again, some idiot is going to push his luck even though the lights are blinking and the railroad guard is down.

The tissues that come together during a swallow to create the seal in the airway have another important trick up their sleeve. They can also flap, or vibrate, when air passes up through them. If the flexibility of the tissues is just right, it can lead to the production of sound. As with so many anatomical features, it was not long until these vocal folds (or vocal cords, as they are more popularly known) were used to impress the opposite sex. Anyone who has lived around a healthy population of frogs knows they are quite capable of significant vocal production. Their abilities are especially noteworthy in the spring when males call repeatedly in an attempt to woo females. Their singing can be downright deafening on warm spring nights—with multiple species creating such a cacophony, it is hard to imagine how the females keep any of it straight. Climb up the evolutionary tree from frogs to humans and both sexes use their vocal abilities to attract mates. In fact, as we'll see shortly, the unique vocal skills of humans are a big piece of the choking story.

THE EPIGLOTTAL GATEKEEPER

There was one final element of the voice box still to come in mammals. The cartilaginous, flap-like epiglottis provided the last blinking sign at the railroad crossing. It acts like a toilet seat that flaps over the larynx during swallowing. For the vast majority of mammals, the epiglottis abuts the soft palate (the back part of the roof of the mouth) and allows for separation of the aero and digestive pathways. It effectively makes the air passageway into an independent snorkel, which allows air to pass to the lungs and circumvent the intersection with the esophagus. This was an incredibly important leap forward for mammals. Because of the epiglottis, nonhuman mammals are terrifically unlikely to choke on their food. It can happen, but it does not happen with anywhere near the regularity seen in humans.

The epiglottal seal with the soft palate also provides nonhuman mammals the ability to eat and breathe at the same time. Simultaneous eating and breathing is a big benefit for furry animals because many mammals spend a good portion of the day eating. Herbivores, in particular, often feed on nutrient-poor food, forcing them to eat a ton of plant matter to meet their dietary needs. The next time you drive by a field of cows or sheep, take note that the bulk of the animals are munching away with their heads down. If a herbivore has its head down all day, it cannot constantly scan the horizon for potential predators. The epiglottis allows herbivores to feed all day and continuously smell the air for predators like wolves, lions, and tigers, who enjoy nothing more than springing sneak attacks on unsuspecting plant eaters.

The epiglottis pushing up against the palate also allows mammals to truly be mammals. That is, it allows them to drink milk from mammary glands. Infants are able to take full advantage of mom's mammaries as the result of the unique anatomical arrangement of the larynx. Because of the epiglottis, mammal young, this time *including* humans, are able to feed from their mothers and continually breathe through their noses. When the timing all works out, a baby can suck, breathe, and swallow in a coordinated manner without having to come off the nipple. This coordination works better in some babies than others, and if they get a runny nose or become stuffed up, as babies tend to do, all bets are off. Then you end up with a sputtering baby trying to get the air and the food in through the same opening, which is not comfortable for anyone.

POLLY WANT A CRACKER

So if the epiglottis makes this nice little seal with the palate, creating effectively separate pathways for air and food, what is the problem? Why is choking a leading cause of accidental death? The answer is that adult humans went and moved their voice boxes. This brings us back to the journal *Dysphagia* for a comparison of the position of the larynx across a range of mammalian species, including humans.[11] I will boil the article down to six words for you: the larynx is lower in humans. The larynx and all of its attendant structures are well down in the throat of an adult human. In a nonhuman mammal (or a human infant), the larynx is basically positioned right behind the tongue.

The features of the descended larynx are still effective at guarding the entrance to the lower airway. With a descended larynx, however, the unguarded upper portion of the airway can become clogged up. The bulk of choking incidents occur when something becomes lodged in the upper airway, above the safeguard features of the larynx. It's as if there are still lights and railway guards present, but they are no longer located where it would make the most sense for preventing accidents.

The low position of the noninfant human larynx does come with a benefit, and it is a pretty big benefit: it allows humans to speak. As air passes up through the vocal cords of an animal, it generates sound. There has to be an adequate amount of vertical space *above* the larynx to shape the sound into something meaningful. Without adequate space, it is not possible to get anything out other than a howl, mew, or bark.

The vocal tract has two sections. Researchers refer to it as the supralaryngeal vocal tract, or SVT. The horizontal portion of the SVT forms a straight line from the lips, through the mouth, stopping at the back of the tongue. The vertical portion is a straight line from that point (behind the tongue) down to the larynx. An adult human has a 1:1 horizontal-to-vertical SVT ratio. A nonhuman mammal has its larynx positioned so high in the throat there is little to no vertical space above its voice box. Many mammals also have elongated snouts, which increases the length of the horizontal dimension. Even if the world's smartest dog, for example, had a lower larynx, it still would not be able to talk because a dog's snout is so long that the ratio would not begin to approach 1:1. If some mad scientist ever decides to use artificial selection to try to generate talking dogs, they should start with a breed that already has a flattened snout, like an English bulldog or a pug.*

I know what you're thinking: so what's the deal with parrots and other birds that can talk? Most birds have vocal organs that actually sit *lower* than the human larynx. A bird's vocal organ, called the syrinx, sits down at the level where the trachea divides into the two primary bronchi. The inability of most birds to talk demonstrates that having a low larynx (or syrinx) is necessary, but not sufficient, for the production of speech. Apparently, what makes parrots and a few other birds unique is their ability to move their tongues in ways that allow them to make vowel-like sounds.[12] This trait is limited to a few select birds (and humans) and is used to create a wide variety of calls in their natural environments. Combined with their low syrinxes, it is their unique tongues that give parrots the ability to say, "Polly want a cracker," or to swear like a sailor, if that's how you like to train your parrot.

Getting back to humans, this is the type of interdisciplinary science where collaboration between biologists and linguists has been necessary to get to the bottom of the story. Such collaborations have led to the understanding that the critical step in the

---

*Even then, they would have to dedicate their entire lives to the project, as we're not talking about fruit flies with two-week generations. Also, the whole experiment hinges on the assumption that dogs are smart enough to talk. But who knows, maybe after 50 years researchers could crank out the first canine that could tell us what it is thinking.

evolution of speech is having enough room to move the tongue around. The tongue needs to be able to dance at the intersection of the horizontal and vertical parts of the vocal pathway. If there is adequate room, as there is in noninfant humans, the tongue can produce dramatic manipulations of the size of the SVT. Without space for the tongue to work, it is not possible to produce the wide range of human consonants and the quantal vowels [i], [u], and [a]. Philip Lieberman, a cognitive scientist at Brown University with a background in phonetics, describes quantal vowels in one of his research articles as the sounds necessary to produce the words "see," "do," and "ma."[13] In all languages those sounds are integral components of speech.

Human infants have an SVT ratio resembling all other mammals. Babies are certainly able to produce significant noise (as all new parents can attest), but they cannot yet shape their hollering into adult-like speech. No matter how many times you stare down at a newborn and make the classic "goo-goo, gah-gah" sounds, they cannot replicate those sounds back to you because their larynx is too high to let them shape words in that way. What they can do is make an adorable "ohhhh" sound over and over again.

Newborns cannot articulate exactly what they want, but they start off life with a degree of choking protection conferred by the epiglottal seal with their palate. The infant epiglottis is so high that it is often visible behind the tongue. My friend who is an otolaryngologist told me he sees "babies in the office all day whose epiglottis is just hanging out in the oropharynx." Oropharynx is fancy doctor-speak for the back of the throat. He sometimes even gets young patients referred to him by primary care physicians who are unaware of this phenomenon and concerned with the white structure they see sticking up in the back of their patients' throats.

In addition to the epiglottal defense, babies feed only on liquids for the first few months of their lives. Liquids are not going to get stuck, even if they go down the wrong tube. Assuming a child is fed only mother's milk or formula for the first few months of life, the only potentially deadly choking hazards are nonfood items.

The larynx starts to descend soon into a child's life. The major drop occurs, typically, between the ages of two and three years old. Given the natural variability in juvenile development, the descent may happen earlier or later for any individual child. The descent continues for a few years, and later other physical changes to the voice box take place with the onset of puberty. The bulk of the laryngeal descent that makes speech possible for humans occurs, however, when we are toddlers. The migration of the larynx creates a problem for small children with their tiny airways. Once the larynx descends, the trachea and esophagus must intersect. The timing of the descent is unfortunate, coinciding with the period in which toddlers are starting to eat increasingly solid food. Just when parents introduce hot dogs and grapes to children, the larynx has dropped into its dangerous, choking-susceptible position.

Ultimately, as with many issues discussed in this book, the story of the anatomy of the human adult throat is one of a trade-off. We have this incredible, unique ability to shape our sounds into meaningful speech. Speech is hugely advantageous and arguably the trait separating us from all other animals. It comes, however, at the cost of having an anatomical arrangement that makes choking a dangerously common occurrence.

FREEDOM OF SPEECH

The benefits of vocal communication in humans are immeasurable and clearly outweigh the costs. Humans have devised many novel systems to facilitate communication (e.g., the written word, Morse code, smoke signals, strapping notes to pigeons, texting), but none of them has ever come close to being as efficient as speech. The other systems invariably demand a greater degree of attention and focus from both the sender and the receiver. It is not easy to engage in another complex activity like, say, weaving a basket or gutting a deer while simultaneously trying to pay attention to or deliver Morse code. In the modern world, many people have hurt themselves, and others, by trying to text and drive a vehicle at the same time.

Human speech is the ultimate tool for multitasking. We do not have to drop everything to engage in conversations with others. The other important feature about spoken communication is that it is a blazingly fast way to get a point across. It is possible to transmit a staggering amount of information in just a few seconds of speech. In addition to the benefit of speed, a speaker with a well-equipped vocabulary is able to communicate almost any idea.

The only rivals in the mammalian world in terms of efficiency and diversity of acoustic communication are cetaceans (whales and dolphins). Cetaceans do not have a low larynx, but whales and dolphins can produce many different kinds of sounds (buzzing, clicking, squeaking, whistling, etc.), and many of them can create multiple sounds at the same time. By varying the volume, frequency, wavelength, and pattern of those sounds, they can communicate a truly diverse array of ideas. It might not surpass the number of ideas humans can communicate, but I would not bring up the issue with a dolphin researcher unless I was ready for a fight or, at least, a lengthy discussion.

So even though the low position of the human larynx makes us vastly more prone to choking, speech stuck around because it is the most flexible and efficient communication system ever hit upon in the long arc of evolutionary history. When exactly the larynx headed south is a difficult question to answer, for the same reason the evolution of the eye is difficult to study: the larynx is made of soft tissue and does not fossilize. The clues left behind are largely in the structure of the skull and vertebral column. By analyzing those structures, researchers have started to piece together the origins of human speech.

The results are somewhat surprising. For much of the evolutionary history of hominins, the face was significantly too long to allow speech to develop. An elongated face meant a much greater horizontal portion of the SVT. Picturing the greatly elongated jaw of a chimpanzee gives the idea. Until the size of the jaw shrunk down a bit, it was not even anatomically conceivable that a hominin could start forming words. This means early members of our genus like *Homo erectus*, living 1 to 2 million years ago, would not have been able to do much more than holler or grunt.

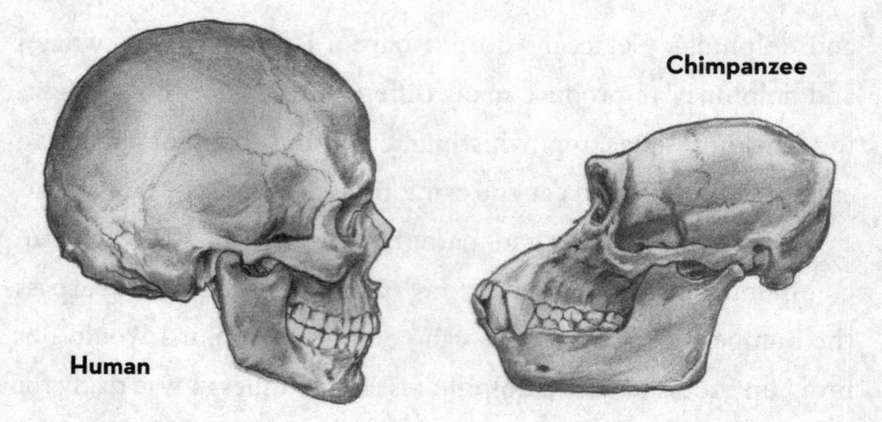

The flat face of humans helps facilitate speech. The chimpanzee face is too long to allow them to form words.

Rolling the clock forward to more recent times, a shorter jaw opened up the possibility of speech. The generally accepted time for the appearance of *Homo sapiens* is approximately 200,000 years ago, though some researchers in the field suggest early *Homo sapiens* may have transitioned away from other hominins by 300,000 years ago.[14] Measurements of skulls and vertebral columns from 100,000 years ago strongly suggest humans from around that time were still not generating anything even close to modern human vocalizations.[15] This means that for at least 50% (and probably closer to 75%) of the time *Homo sapiens* have roamed the earth, they have been incapable of producing the full range of human speech.

The hang-up has to do with the length of the neck. The neck had to elongate to allow the larynx to descend, creating the necessary 1:1 horizontal-to-vertical ratio. Even with a shorter jaw, without elongation of the neck, the larynx had nowhere to descend without causing a suite of other problems. Based on fossilized cervical vertebrae of early humans, the neck does not appear to have elongated to anywhere near its current length until around 50,000 years ago. While the early advantages of a longer neck were likely related to speech, these days many people consider an elongated neck a symbol of beauty and status. Some cultures go so far as to place rings around the necks of young children in an attempt to create the illusion of long and elegant necks.

Some paleoanthropologists think the neck issue may have kept our close cousins, the Neanderthals, from ever swapping tales and singing songs around the campfire. Recent evidence pins down the date of their extinction to roughly 40,000 years ago,[16] and by then, even though their faces were not overly long, they still had squatty necks that may have provided insufficient room above their voice boxes to allow for speech. The Neanderthal research community hotly debates to what degree our close cousins were able to communicate orally. There are some clues in the bones of the head and neck, but most of the story is found in soft tissues now lost to history.

Regardless of whether or not they could talk, Neanderthals must have had some fine features because we now know that Neanderthals and humans interbred. The evidence for this fun fact came from the sequencing of the human and Neanderthal genomes, which demonstrated some distinctly Neanderthal DNA patterns scattered within the genomes of modern humans. Such DNA swapping could have happened only one way and is evidence of the slightly uncomfortable fact that, at some relatively recent point in our history, humans and Neanderthals had sex. If Neanderthals were not capable of articulate human-like speech, maybe early humans occasionally mated with Neanderthals because they were exceptional listeners, unable to respond to much of a degree even if they wanted to.

ANOTHER PROBLEM . . . SLEEP DISRUPTION

Other changes, in addition to the lowering of the larynx, helped facilitate speech in humans. The cranium changed in several ways that led to a narrowing of the throat. For example, the foramen magnum, the large hole in the base of the skull through which the spinal cord passes, gradually migrated forward, ending up much farther forward in humans than in other primates. As space opened up with the descending larynx, the palate and tongue both moved back deeper into the throat. A narrower passageway resulted that was useful for speech because it allowed for refined modulation of the sound made by passing air through the vocal cords. Crafting speech with the earlier, wider opening would have been like trying to make music by blowing into a roll of wrapping paper. With a narrower opening, humans had a built-in musical instrument, something more like a flute, or at least a kazoo.

Those same anatomical changes, however, led to another serious issue that manifests uniquely in humans. I've highlighted choking here as the leading biological cost of a lowered larynx, but the conversation is not complete without at least addressing the issues of sleep and snoring. For many people, the anatomy allowing for speech totally fails them when they lie down and attempt to rest. Sleep-disordered breathing (SDB) is the technical name of the condition, and it is shockingly common. Research published in the *New England Journal of Medicine* suggests the problem occurs in 24% of adult males and 9% of adult females.[17]

In its less severe forms, SDB may only cause an individual to snore. This may not disrupt the sleep of the snorer, but it can have a very negative impact on anyone trying to sleep nearby. The most severe form of SDB is obstructive sleep apnea. Obstructive sleep apnea is a problem limited to humans and to a few breeds of dogs with flattened faces from generations of artificial selection.

For people with sleep apnea, the issue is no joke. Relaxation of the muscles in the throat during sleep drives the disruption. The relaxed muscles lead to a narrowing of the airway and create a problem because the human airway is already uniquely narrow. If we didn't have such long, narrow airways, it wouldn't matter as much if our muscles relaxed during sleep. Obstructive sleep apnea causes sufferers to cease breathing while sleeping. This forces the body awake in order to reset the system, which leads to the interruption of sleep. People with sleep apnea often only partially wake up, so they sometimes do not even know all of this is happening. They just know their quality of sleep must have been garbage because they end up dragging themselves through the next day. Often it is up to their partners to figure it out. The partners are the ones lying awake (thanks to the excessive snoring) listening to the cessation and resumption of the breathing of the afflicted.

If left untreated, obstructive sleep apnea can spiral into a suite of other issues such as insomnia, headaches, and depression. Obviously, it also leads to serious difficulty with concentration and alertness during daylight hours. Sleep apnea is even associated with a wide variety of potentially deadly conditions such as heart attacks and strokes.

Terence Davidson, the author of a research article published in the journal *Sleep Medicine*, explores the link between vocal tract anatomy and disrupted sleep. He asks the natural question of whether there was "adverse selection for sleep apnea." He answers his own question by stating, "Other than an occasional snorer who was killed by his cavemates, most likely there was not negative selection for sleep apnea."[18] I guess the typical snorer in the cave must have just been kicked outside so everyone could get some rest.

Another angle Davidson brings up is that the extreme forms of sleep-disordered breathing, such as obstructive sleep apnea, are much more common in older individuals. The muscles in the throat (and everywhere else, for that matter) naturally start to lose some of their tone with age. The age factor means natural selection has not had the opportunity to shape the anatomy in ways to prevent the problem. Thousands of years ago, most people would have been long dead before they reached the ripe old age of 40, when SDB starts to surface in much of the population. Even now, most people have children before problems with SDB typically begin. As a result, severe conditions like obstructive sleep apnea avoid the filter of natural selection and keep getting kicked (or snored) down to the next generation. The only way natural selection is going to take care of the predicament is if we all agree to never have sex with people who snore, which seems both unlikely and unfair to the snoring population. Even then, the snorers would probably all just find other snorers to have sex with, which would only make the problem worse by generating a new lineage of super-snorers.

THE LONE TALKERS

If the benefits of oral communication can outweigh the costs of choking and sleep disruption, why has it not become more widespread in animals? One hypothesis suggests speech never would have occurred if human ancestors had continued running around on all fours. Quadrupeds are highly dependent on their sense of smell; they take in much of their information about the world through their noses. We intuitively understand this by casual comparison of our sense of smell with basically any other mammal. Flatten out the olfactory epithelium from a human and it makes a piece of tissue about the size of a postage stamp. Do the same thing for a dog and the piece of tissue is more like the size of two smartphone screens.[19] The comparison is apt, because if you think of the human nose as working at the level of snail mail, dogs have the latest iPhone built into their snouts.

Picture how a dog walks down the street, sniffing every flower and fire hydrant it encounters. When a dog comes across another dog, instead of making eye contact and shaking hands, the two dogs sniff each other all over (some of them really get in there) and, in doing so, probably gather more information than we can even imagine. Because of the anatomy of a dog's head and neck, they are able to take in a stream of olfactory information without interruption. For many animals, interrupting their sense of smell would be akin to putting a blindfold on a human or having every blink last a few seconds.

When human ancestors became bipedal, the argument goes, it started a gradual shift away from the sense of smell as the dominant sense. With the new upright position, hominins became more dependent on their sense of vision. Being able to see potential trouble, rather than smell potential trouble, hominins did not need to constantly sniff the air for predators, and mutations that altered the shape of the throat had a chance to stick within populations. The tight lockup between the epiglottis and soft palate loosened in early hominins, and the larynx began its descent.

Eventually the shape of the throat changed significantly enough to allow humans to do more than just say "ugh, ugh" around the campfire. They weren't exactly singing "Ave Maria" just yet, but whatever they were doing was clearly a fair bit sexier than simply grunting. The selection for vocal production rolled on, and eventually we ended up with a throat that allows us to sing karaoke but also makes it tricky to sleep and a little bit scary to eat a grape.

PART TWO

Achy-Breaky Parts

4

Two Snakes a Day

The world record for a human running the 100-meter dash on two feet is 9.58 seconds. What is the world record for the 100-meter dash for a human *running on all fours*?

a. 8.32 seconds

b. 15.71 seconds

c. 22.93 seconds

d. 45.03 seconds

I was a super skinny kid. I was five foot six (168 cm) and 95 pounds (43 kg) when I got my first driver's license. To this day I wonder why, just for my own self-esteem, I didn't fudge the weight a little to at least get up to triple digits. I also played a ton of different sports while growing up. The combination of being stick thin and very active led to a fair number of broken bones. I was quite accomplished at crashing into fences, defenders, and all other manner of obstacles. I always knew right away when I broke a bone. The level of pain was different from other injuries. All these years later I still have the impression that if you wonder *if* you have broken a bone, then you have *not* broken a bone. When you break a bone, your body tells you loud and clear. You can sometimes play through the pain of a bruise or a sprain because the pain is less intense. The pain of a broken bone is sharp and unyielding, and there is zero question you should stop moving the joint and get some help.

I eventually filled out a bit and stopped having so many trips to the ER. I had a long injury-free run until a few years ago when I was a faculty member playing on a basketball team with some colleagues and the injury bug bit again. Most of the other teams in the league were made up of students. That was clearly the first mistake right there. It was probably not the best idea to get up from my desk, warm up for about two minutes, and attempt to chase a bunch of young college students around a basketball court.

I remember being down low by the basket and putting up a shot fake. The defender bit, but instead of going up to score, I crumpled to the floor. I knew immediately I had torn my anterior cruciate ligament, or ACL. In my years of playing basketball I had experienced countless sprained ankles and knees, stretching and straining ligaments and tendons but never severing them in two. This pain was different. It was more akin to the pain I remembered from breaking a bone.

I imagine having an ACL repaired is an interesting experience for anyone, but it is especially so for an anatomy professor. After years of teaching about the structure of the knee, I got to experience the fragility of its anatomy firsthand. I was lucky with my injury and avoided any collateral damage. An ACL tear often comes with the added bonus of a torn medial collateral ligament (MCL) and a torn medial meniscus, a situation known as the "unhappy triad." I happily avoided the unhappy triad and had the relatively straightforward scenario of needing to have only one ligament repaired.

The most common first step in an ACL repair is to donate a piece of your own body to use in the surgery, a rob-Peter-to-pay-Paul scenario. The two most popular repair techniques involve harvesting either part of the patellar tendon (the thick piece of tissue below the kneecap) or a tendon from one of the hamstring muscles that run along the back of the leg. The orthopedic surgeon then uses the harvested tissue to create a new ACL.

You do not have to get a torn ACL repaired. The ACL provides a great deal of stability to the knee, especially for movements like twisting and pivoting, but you can survive fine without one. I had a friend in graduate school who tore his ACL, also while playing basketball. He chose not to get it repaired and carried on with a ripped-apart knee. After the swelling subsided and the pain went away, his day-to-day life was not notably different with a torn ligament. It's probably harder for him to juke a defender with a crossover dribble, but you can very effectively get through life without a deadly crossover. It is now nearly 20 years later and he still has not had it repaired. In fact, before surgical techniques became more refined and effective, it was not unheard of for professional athletes to carry on with their careers in spite of torn-up knees. Several athletes, including Mickey Mantle and Joe

Namath, are believed to have played large stretches of their careers with torn ACLs.

I wanted to be able to move confidently on the court in the future, so I read exhaustively about the two repair techniques and chose a surgeon who reconstructed ACLs with hamstring tendon grafts. The hamstring technique usually involves a slightly longer rehab, but it also brings a decreased likelihood of residual knee pain. That trade-off sounded good enough to me. On the day of my surgery, I made sure they marked the correct leg to cut open,[1*] held the anesthesia mask over my face, started to count down from 100, and was out cold before I got to 90.

A few weeks later I was back at the surgeon's office for my first follow-up appointment. I had several questions about the rehab schedule and also wanted to pick the doctor's brain about how the surgery had gone. He made some comments that have always stuck with me. He described the notch in my knee into which the ACL fit as a "little A-frame" and did not seem the least bit shocked that the ligament had torn. If anything, given all the sports I played, he was surprised it had lasted as long as it did and expected I would blow out my other knee at some point. He implied that my joint was set up for failure by the size and shape of my ACL. One overzealous pump fake had finally detonated the ticking-time-bomb anatomy of my knee.

---

*Wrong-side surgical mix-ups are surprisingly common. One study focused on surgeons in Colorado over a 6.5-year period and found 107 incidents where the wrong side was operated on and another 25 cases where the wrong *person* was operated on.

And thus, with my blown-out ACL, we transition to a whole new suite of problems for the human body—the aches and pains of the skeletal system. As with the issues of the head and neck, the story behind our achy bones, cartilage, ligaments, and tendons lies deep in our past. We'll look at problematic anatomical hot spots throughout the body in this section, but we might as well start with the joint that puts more people under the knife than any other—the knee.

THE TIP OF THE ICEBERG

My injury was far from unique. Millions of people tear their ACLs each year.* Most ACL injuries occur in the context of playing a sport, but the majority of the injuries, like mine, happen in a noncontact manner rather than, say, from a violent tackle. The most common threat comes from landing awkwardly after jumping or planting a foot to make a cut or a pivot. Most people don't tear their ACLs running in a straight line. ACL injuries most commonly happen in sports like soccer, football, and basketball in which players spend the whole game jumping or firmly planting their legs to quickly change direction.

A life of avoiding sports with twisting movements does not automatically guarantee freedom from knee pain and injury. Ligaments like the ACL and MCL are most commonly torn through

*Based on numbers from the CDC, 250,000 people suffer an ACL injury each year in the US alone.

acute and traumatic sporting-related injuries, but other soft tissues in the knee can be injured in a more benign way. The menisci lie between the femur (the thigh bone) and the tibia (the large bone in the lower leg). They act as shock absorbers and are particularly injury-prone. As with ligaments, menisci are vulnerable during motions of twisting, but they do not require the typical full-speed nature of a ligament injury to tear.

One study published in the journal *The Knee* went through the records of every single patient who underwent arthroscopy for suspected meniscal injuries with a particular surgeon over a period of eight years.[2] The existence of an entire journal dedicated to one joint does seem to suggest it is a troublesome anatomical feature. The authors focused their study on people who injured their knees for the first time. Eliminating the repeat offenders whittled down the subject list from 1,236 patients to 392. More than two-thirds of the patients who came through the doors of the clinic were people the surgeon had seen in the past for knee problems. Clearly knee injuries beget knee injuries.

Nearly one-third (113) of the 392 first-time patients could not qualify their injuries as sporting or non-sporting. They just had painful knees, did not know why they had become painful, and wanted them fixed. In other words, many of the patients injured their menisci simply going through the daily paces of life. They had not hurt their legs in weekend pickup games or as a result of some slip on a curb. Their menisci wore out because sometimes menisci simply wear out.

The study shined a spotlight on the nature of meniscal injuries and, in particular, stressed how they commonly occur independently of sports. Only 127 of the total 392 meniscal injuries (32.4%) were sports-related. In 152 of the cases, the patients could clearly identify the nature of their injury as non-sporting. Interestingly, a significant trauma was not the cause of the majority of the non-sporting meniscus injuries. Sure, there were a few grisly motorcycle accidents or nasty falls down the stairs mixed in, but the most common cause of injury in the non-sporting group was crouching. Yes, crouching.

The simple process of bending down to pick something up can lead to a tear in the meniscus. People with a meniscus tear often end up going under the knife because the body is lousy at self-repair of the knee, for reasons we'll explore later in the chapter. The elderly are not the only victims of crouching injuries. Squatting was the "most frequent cause of a major meniscal tear in patients less than 30 years in the non-sporting injuries group."[3] I guess that makes sense. Most of my well-intended but ultimately foolish squatting (e.g., moving a heavy couch, picking up a girl) occurred in my 20s. By the time you get older, you pay someone in their 20s to move your couch, and you are sort of beyond the stage of physically picking up women to impress them.

LIFE AT THE TRANSITION

Let me spoil the ending for you: It's because we're bipedal. Our knees are injury-prone because we walk around on two feet. Evolutionarily speaking, the transition from moving around using four limbs to moving around on two limbs just did not happen all that long ago. There is general agreement in the scientific community about when it occurred. Some of the earliest direct evidence comes from fossilized footprints in Laetoli, Tanzania. Members of *Australopithecus afarensis* (Lucy's species) made footprints there 3.6 million years ago.[4] Other evidence, from the remains of the *Ardipithecus ramidus* fossil, Ardi, push the date back to 4.4 million years ago.

That might seem like a long time, but 3 or 4 million years is a mere blink in the overall timeline of life on Earth. It is the last page of an exceptionally long book. If you condense the whole timeline of the universe into one calendar year (as Carl Sagan famously did in *Cosmos*), 4 million years is the last few drunken hours on New Year's Eve. The eye, for some perspective, has had 375 million years to sort itself out in the transition from water to land. That's nearly 100 times as much time. Four million years into life on land and I bet the view of the world through the eyes of a proto-amphibian was still relatively blurry. The human knee is still working through the kinks in the early stages of its transition.

It is not exactly a trivial switch going from walking on four limbs to two. Some special hominid (the taxonomic family that includes humans and our extinct close relatives and ancestors *and* the other great apes) did not simply come out of the trees and decide to run a few miles on two feet. There had to be a transitional period. Scientists have generated as many as 30 hypotheses that attempt to explain how and why the transition to bipedalism took place. Charles Darwin himself kicked off the bipedalism hypothesis party in 1871 in *The Descent of Man* (his second most famous book after *On the Origin of Species*), in which he described an idea later branded as the freeing-of-the-hands hypothesis. Many other authors jumped on the bandwagon in the 20th century and refined the scope with ruminations about all the advantages free hands would have allowed in regard to tool and weapon use.[5]

The whole point of scientific hypotheses is to subject them to experimentation to see if they hold up. A hypothesis cannot be proven true. Scientists gather evidence, and the evidence either supports or does not support the hypothesis. If there is a mountain of evidence in support of a hypothesis, the language changes to use terms like "theory" or "law" to describe the idea (think theory of evolution or theory/law of gravity or the atomic theory). In other instances, evidence leads to the rejection of a hypothesis. If the idea never proceeds through the steps of experimentation and evidence gathering, then it does not become grounded in science.

Many aspects of the freeing-of-the-hands hypothesis ran into trouble once the evidence surfaced and the timeline of human evolution became clearer. If we conservatively place the roots of bipedalism somewhere in the range of 4 to 5 million years ago (some argue it started even earlier), that leaves at least a couple of million years before the first indications of primitive tool use appear around 2 to 3 million years ago. So selection eventually occurred for behaviors like chucking spears and swinging hammers, but tool use does not explain why bipedalism occurred in the first place.

By the late 20th century there were some new ideas. The thermoregulation hypothesis suggested the main benefit to bipedalism was a significantly reduced cost compared with quadrupedal life in regard to exposure to solar radiation. In more modern terms, it's the difference between standing in line for the water slide and sunning yourself by the pool on a lounge chair. You're more likely to overheat and get burned relaxing by the pool because you have a much greater surface area exposed to the sun. Quadrupeds stand in a position with a high degree of sun exposure. Prop 'em up on two legs and the system is less likely to overheat, or so the thinking went.

This hypothesis makes a lot of sense, assuming the transition took place in an open savannah environment, where thermoregulation would have been a critical challenge to overcome. As more and more fossil evidence was unearthed in the late 20th century, however, it became clear that assumption was also wrong. The earliest hominin fossils are from forested environments, which were likely wet and humid.[6] This revelation was a death knell for the thermoregulation hypothesis. Like tool use, a cooler body was likely a later benefit accrued by bipedal hominins but not part of the original impetus to migrate away from a quadrupedal life.

THE WADING APE

A 21st-century hypothesis takes this new knowledge about the wet environment into account and has presented a radically different idea for how the transition took place. The amphibian generalist theory (which is more of a hypothesis than a theory, in my opinion) suggests the critical behavior in the evolution of bipedalism was wading. Its authors suggest early hominins descended from trees into marshes and areas of low-lying water in order to forage for plants and small animals. Around 6 to 7 million years ago, the climate in Africa became highly variable.[7] In extremely wet periods, with an increasing number of marshes and lakes, wading apes would have been forced into a bipedal stance in order to keep their heads above water. While wading, foraging in the marshes with free hands would have opened up new dinner options. It turns out Darwin may have been right about the utility of free hands, even if he missed the mark about their initial use.

This newer idea also meshes well with observations of the bipedal gait of modern great apes. Other great apes waddle and have to flex their legs to adopt a bipedal stance. Unlike humans, they are not able to lock their knees and stand up perfectly straight in a relaxed way. As a result, chimps and gorillas fatigue very quickly when walking or standing bipedally *on land*.

You don't have to take my word for it. Adopt your best bipedal chimpanzee pose and see how long you last standing there with your legs slightly flexed. When I do this, my knees start hurting after about 30 seconds and then some back pain kicks in after only a minute. Two minutes of this position and I become downright miserable. I tried it for two minutes a couple of hours ago and I can still feel the pain in my back. It was probably not the smartest way to illustrate the point, but curiosity got the best of me.

Introduce some water into the equation and it changes everything. Chimpanzees and gorillas naturally adopt a bipedal stance in shallow water. Their behavior appears strikingly similar to how humans walk through hip- and chest-deep water, complete with raising their hands above their heads if the water gets deep enough. The whole key to the amphibian generalist theory is that the buoyancy of the water relieves the strain of the quadriceps in a flexed-leg biped and helps support the weight of the body.

After trying out the flexed bipedal ape stance on land (and finding it difficult, to say the least), I went down to the local pool to try it out in water. The difference is incredible. I feel like I could stand in an ape crouch all day long in hip-deep water. Early hominins would have lasted only a few minutes standing on two feet on land, but they would have been able to stand in water for hours on end. The support of the water nearly completely removes the strain of the muscles.

Once you ape-crouch in hip-deep water, it is not difficult to imagine how water may have been the necessary bridge to bipedalism. It could have supported the hominin body through the transition while allowing for exploitation of new environments. In one of his research articles on the subject, German anatomist Carsten Niemitz, one of the biggest champions of this wading hypothesis, argues that bipedalism via wading is the "only scenario suitable to overcome the considerable anatomical and functional threshold from quadrupedalism to bipedalism."[8] After my experience crouching as a great ape on land and in the pool, I tend to think the amphibian generalist theory may just hold water.

An even newer idea links the origins of bipedalism with sex. As the climate changed and food became harder and harder to come by, the food shortage would have been an especially pressing issue for females because the significant responsibilities of child-rearing limited their ability to forage over long distances. Under such a scenario, there would have been strong selection for males able to provide for females. The only way a male could forage and bring food back would have been to adopt a bipedal stance and carry his bounty in his arms. This is a difficult hypothesis to test, but there is some modern-day evidence in support of the idea: chimpanzees switch to bipedalism when carrying particularly valuable resources.

The author of this walk-for-sex hypothesis, Owen Lovejoy of Kent State University, suggests this was the historical moment when monogamous behavior entered the hominin line.[9] The reward for a provider male may have been a faithful female back at camp. With bipedalism and long-distance foraging, males may have shifted away from beating their chests and baring their teeth to impress females to figuring out which type of food gifts were the perfect ones for getting potential mates in the mood. A few million years later and males are still trying to figure that one out. In my experience, potato and leek soup with chopped chives and crumbled bacon gives you a good chance.

CHIMPS ON TREADMILLS

Even with solid evidence supporting some of these hypotheses, we will likely never know with 100% certainty why human ancestors became bipedal. Whether it was because of rain, sleet, snow, heat, cold, sex, foraging, wading, carrying kids, carrying food, making fire, throwing rocks, catching rocks, chucking spears at fire-breathing alien dragons, or all of the above, what matters most for our anatomy is simply *that* it happened. As such, we now shift our focus to the adaptations of the human body that make bipedalism possible.

They say you have to walk before you run, and this was also the case for our ancestors. In the first few million years of bipedal life, several significant anatomical changes occurred that made life on two feet possible. The skeleton required some serious remodeling if hominins were going to stand on land for any period of time without wanting to holler out in pain. One of the most obvious skeletal differences between apes and hominins is the angle of the femurs from the hips to the knees. In a chimpanzee, the femurs are nearly vertical, whereas in early hominins and humans the femurs angle inward. This has the effect of placing the knees and feet closer together in humans, below the center of gravity of the body, which makes for an efficient stride with less shifting of body weight (we don't have to waddle like chimps) and helps with the not-insignificant challenge of balance when walking on two feet.

The pelvis also underwent a significant remodeling with the switch to bipedalism. The modifications helped with the transition to a bipedal stride, but the shape of the birth canal changed as a result. The issue is covered in a later chapter, but long story short, the changes made for a serious, even life-threatening, issue for human females when trying to birth children through such a tight space.

Another major shift occurred with the shape of the spine. The quadrupedal great ape spine is C shaped, with the center of gravity positioned far forward. Because of the shape of their spines, when walking bipedally, chimps have to work hard not to tip over. It is yet another reason they are unable to stay upright for long without growing tired. In contrast, the S-shaped human vertebral column puts both the head and the torso above the center of gravity, instead of in front of it. As with the other changes, the switch

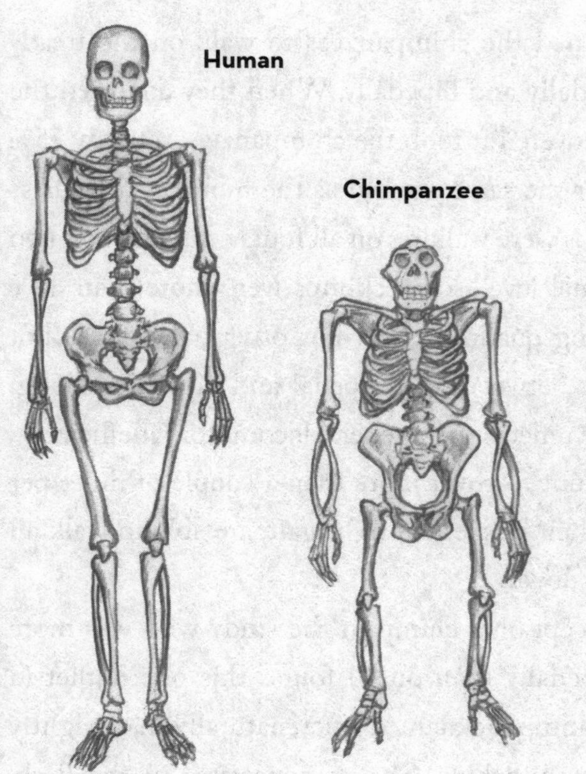

Human femurs are angled inward, which works great for standing. Chimpanzee femurs are much more vertical, which is lousy for standing.

in spinal curvature came with significant anatomical compromise. The human epidemic of back pain is the cost this time, and it is a cost we'll also revisit later in more detail.

These changes and others have made humans very efficient at walking. To understand how efficient, for comparison's sake, researchers at the University of Arizona trained chimpanzees to walk on treadmills while hooked up to equipment measuring their oxygen use.[10] (This is only the second-oddest animals-on-treadmills story I know, as I have a friend who once coaxed vampire bats into running on tiny custom-made bat treadmills.)*

*Search Dan Riskin vampire bat treadmill in Google Videos to check out bats running on treadmills.

The scientists trained the chimpanzees to walk on the treadmills both quadrupedally and bipedally. When they analyzed the group data, they discovered it took the chimpanzees roughly 75% more energy to cover the same ground as the humans, regardless of whether the chimps were walking on all fours or upright on two feet. On the individual level, some chimps were more than 30% more efficient walking quadrupedally. For others, there was not much of a difference between quadrupedal and bipedal walking. Chimpanzee lifestyle reflects their general locomotory inefficiency, in that chimps tend not to cover more than a couple of miles per day. Humans, with their relaxed, upright gait, are able to walk all day without wearing down.

There was one exceptional chimp in the study who was more efficient walking bipedally than on all fours. This one outlier (a 33-year-old female chimp nicknamed Lucy, naturally) had slightly more extended knees and hips. She demonstrates in the flesh, rather than through the study of fossils, how there can be enough skeletal and muscular variability present in a quadrupedal animal to allow for selection of traits favoring bipedalism. A chimp like Lucy might just survive if the jungle flooded and chimps had to walk upright.

The researchers did not, unfortunately, make their human subjects also walk on all fours. I would love to have seen the variability in quadrupedal efficiency in humans. Believe it or not, there is an established human world record for the fastest 100 meters running on all fours. The Usain Bolt of four-limb running is Kenichi Ito from Japan, who holds the world record with a time of 15.71

seconds (compared with Bolt's 9.58 seconds running bipedally). Just as it is somewhat striking to see a nonhuman great ape amble around on two legs, it makes you do a double take to see a human sprinting down the track using all four limbs.* It also stresses, in a very visual way, the notion that our quadrupedal past is, in fact, not so very deep in the past.

THE FLIP-PHONE MENISCUS

The switch to bipedalism presented the knee with a variety of challenges. Perhaps the most obvious is weight distribution. Four sets of shock absorbers nicely distribute the weight in a quadruped, but in a biped only two sets of padding bear the burden. Having to do double duty, those shock absorbers wear out and frequently tear. The hominin knee became quite bulky in the few million years following the first bipedal steps, as an increased surface area helped to handle the increased workload.

The shape of the pads also changed as natural selection pushed the anatomy in a direction favoring transportation on two feet. There are two of the shock-absorbing menisci in each knee, a lateral one toward the outside of the knee and a medial one toward the inside. The medial meniscus has not changed much in the recent evolution of humans. It is crescent shaped, has two insertions on the tibia, and effectively resembles the medial meniscus of all other primates.

---

*Searching 100 meters on all fours in YouTube pulls up videos of Kenichi Ito setting the record.

The lateral meniscus is where there is significant variability across species. There are three basic arrangements of the lateral meniscus in primates. In all of the platyrrhines (the new-world monkeys), the lateral meniscus is crescent shaped with one insertion on the tibia. In all nonhuman catarrhines (the old-world monkeys and apes), it is ring shaped with one insertion—with the exception of orangutans, which have knees set up like the new-world monkeys, for some reason. (If I were an orangutan writing this book, I would want to know why orangutan knees are structured like howler monkey knees instead of ape knees. But as a human, I'm more interested in exploring the unique human lateral meniscus. The orangutans can sort out their own evolutionary quandaries.)

Humans are the only primates with crescent-shaped lateral menisci with two insertions on the tibia. There is a 407-page book dedicated exclusively to the meniscus (its title, obviously, is *The Meniscus*), and in its first chapter the authors describe the second tibial insertion as helping to support the knee during the "stance and swing phases of bipedal walking."[11]

Most humans have crescent-shaped lateral menisci. All these decades of surgeons carving open people's knees, however, have revealed a surprisingly high number of people with disc-shaped (or discoid) lateral menisci.

Menisci develop very early (they are present as early as eight weeks into gestation), and if they develop normally, at no point in the process are they disc shaped. A discoid lateral meniscus is considered an atavistic character. An atavistic trait develops as a reversion to the ancestral form of the character rather than the modern form. This is a different phenomenon than the presence

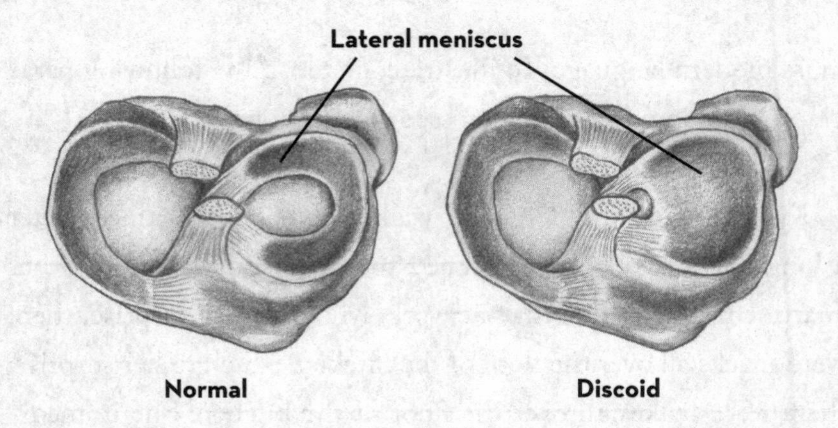

Most humans have crescent-shaped lateral menisci. Unlucky humans have one or both lateral menisci shaped like discs. The discoid version is much more injury-prone.

of wisdom teeth or an appendix. Evolutionary compromise forced those structures into awkward and imperfect situations, but they are nearly universally present. Atavistic characters are rare throwback traits, present only in individuals whose bodies, for whatever reason, forgot the time and age they were living in. It's like a brand-new car with a tape deck in it, or like seeing a stylish college student pull out a flip phone. Everything else is set in the modern time except for one jolting atavistic feature.

SNAKES AND SNACKS

Another way to think of atavistic traits is to view the genome as a living, written document, but one in which all the earlier drafts of the story are still present in the book. Instead of permanently deleting all the old text when edits are made or new text is written, the body has kept many of the old passages sprinkled among the more modern language. To illustrate, first read the following paragraph, skipping over all the crossed-out sections:

Kylee opened the door and walked into her apartment after a long day at work. ~~Kylie opened the door and walked into her apartment after a long day at work.~~ Much to her surprise, there were snacks all over the floor of the kitchen. ~~Much to her surprise, there were snakes all over the floor of the kitchen.~~ She immediately thought about her kids and quickly went off to find them. ~~She immediately thought about her kids and quickly went off to find them.~~

The crossed-out sentences represent genes that are still present in the genome but which are no longer active. Now go back and read the same paragraph again, and this time skip the regular text and instead read the three crossed-out sentences.

In this short paragraph about a messy kitchen, there are three repeated sentences (or genes). In the first sentence, the older version is effectively the same as the newer version. You might not have even noticed the change. The modern version spelled Kylee ending with "ee," and the older version spelled it ending in "ie." Such a mutation is known as a silent mutation, and when a silent mutation pops up in DNA it causes no change to the actual trait. The change is effectively silent.

The edit in the second sentence is also minor (requiring only two mutations), but this time the change is significant. In one version, you have the innocuous outcome of some crackers spilled on the floor. In the other version, there are snakes all over the place! The difference is dramatic.

In the last sentence, the old and new versions are 100% identical. There is just an extra copy of the information lying dormant in the paragraph. This kind of thing is dirt common, with sometimes dozens of copies of genes littered throughout the genome.

I picked up a brochure with a memorable typo a few years ago at a day-care facility, and it inspired this snack/snake story. Under the section about meals they had written, "Each child will be provided two snakes a day." The brochure did not provide any additional information about what types of snakes the children would be given. Obviously, you would want it to be something with a little meat on the bone, and not just a little garter snake or something.

Given that the human genome contains three billion pieces of code, each cell has to successfully piece together a story from a document containing three billion "letters." That is a lot of "letters" to read without making a single mistake. I sometimes challenge myself at night when reading bedtime stories to my daughter to get through an entire book without making a single error. It is not as easy as it sounds, and as the books get longer it only gets more difficult. Printed out single-sided on regular copy paper in Times New Roman 12-point font and with one-inch margins, all the letters of the human story, or genome, would take up 1,206,980 sheets of paper.* That is one doorstop of a story.

---

*A professor at the University of Minnesota did the math. Google Michael Miller stack of paper genome to see the calculations.

Every cell of the body, with the exception of mature red blood cells, contains the entire story, all three billion letters of it. Most of the time, cells read the most recent edition and use only the most updated versions of genes. But every now and again, cells use one of the previously crossed-out genes, and you end up with snakes in the kitchen. Or, to pivot to a real example, sometimes the body goes with an older variant and a baby is born with a tail. For real. Every human has a short tail early on during embryonic development, but the tail typically shrinks and disappears before birth. Once every blue moon, however, during fetal development, the body fires up some decidedly old-school genes, the tail does not reduce, and a human is born with a true tail. It is usually not overly long, but even a tail that is only a few centimeters long looks rather striking on a human.[12] Go ahead and type "caudal appendage" into Google Images to see for yourself. It does engender one of my favorite cocktail party questions: "If you had to have one of the two, would you rather have a short tail or webbed feet?" You could hide a tail, but the webbed feet would be great for swimming. I go back and forth but am mostly happy to just have to think about it in hypothetical terms.

Not all atavistic characters are super rare. Roughly 5% of people have at least one extra nipple. This classic atavistic trait, called a supernumerary nipple, appears in around the same percentage of the population that gets stuck with a discoid lateral meniscus.[13,14] In the ranks of conversation starters, a discoid lateral meniscus ranks far below a tail or an extra nipple, but it is much more likely to cause problems in the long run. Tails are typically removed surgically after birth with no ill effects, and even a few extra nipples are not usually a problem. I suppose they could even come in handy in the case of triplets because, yes, for a lucky few, sometimes those supernumerary nipples are even functional.

SHIFTY KNEES

I could write about human tails and extra nipples all day, but getting back to knees and why they give people so much trouble, an atavistic discoid lateral meniscus is typically missing its posterior attachment to the tibia, and it is usually thicker than the normal crescent-shaped variety. Thicker doesn't sound bad on the surface (the whole point is shock absorption, after all), but the thicker discoid version is more prone to tearing than the regular version, and a torn meniscus usually means surgery.

One of the more striking observations about knee surgeries, and joint surgeries in general, is the overall lack of blood. I show videos of several surgeries in my anatomy classes. I figure all these future doctors and nurses better figure out if they are okay with blood before they get too far down the medical path. No one ever turns green watching an ACL or meniscus repair. There's almost no blood throughout the entire surgery. (Fire up a video of a quadruple bypass or prostate removal on the big screen and it's a whole other ball of wax. The surgeons spend half their time cauterizing vessels and sucking up blood, and inevitably, a few of my students rethink their career plans. They quickly figure out that computer science or business administration involves far less blood and pus.)

Blood brings in all the new materials the body uses to repair tissues. Repairing a knee is like trying to fix up a house in a very rural location. Without significant blood flow, the biggest challenge is getting the raw materials on site. The knee cannot do it, which is why a torn ACL stays torn forever unless there is surgical intervention. The situation is very different with, say, a broken bone, where the body can often repair a break as long as it is kept immobile for a long enough period of time. A broken bone is a house located right next to an interstate, making it easy to bring in new shingles and sheetrock. Bones have blood coursing through them, so it's a cinch to access the necessary materials to repair a break.

The inability of knees and other joints to self-repair is why we have orthopedic surgeons. With less refined surgical techniques, old-school treatment of torn menisci often involved a complete meniscectomy. Total removal of the meniscus can work in the short term thanks to the abundance of other cartilage in the knee. Even if the lateral meniscus is fully removed, there is still some shock absorption achieved by the medial meniscus and the articular cartilage present at the ends of the bones. In the long term, however, a meniscectomy puts patients on the path to a diagnosis of early-onset arthritis. In contrast, newer techniques attempt to leave as much meniscus intact as possible. In the case of the discoid lateral meniscus, some surgeons now go so far as to carve the discoid meniscus into the proper crescent shape. The process is called saucerization, which is my new favorite medical term. What?! Your body forgot you were bipedal and dealt you a primitive meniscus? No problem, we'll just saucerize that sucker and make it human!

This is the point in the story where the knee injuries start to compound. If you will recall, most people through the door of an orthopedic clinic are repeat customers. The torn meniscus is often just the beginning. Once a knee becomes unstable, the injuries can start to pile up. This is especially true when someone draws the unlucky card of a discoid lateral meniscus. The authors of a research article with the short title "The Discoid Meniscus" state, "Discoid lateral meniscus was reported to be associated with other musculoskeletal anomalies, among them high fibular head, fibular muscular defects, hypoplasia of the lateral femoral condyle with lateral joint-space widening, hypoplasia of the lateral tibial spine, abnormally shaped lateral malleolus of the ankle, and enlarged inferior lateral geniculate artery."[15] Yikes.

Let me translate. Discoid lateral meniscus is associated with a seriously jacked-up knee, a funky-looking ankle, and some blood-flow abnormalities. It's tough enough to get your knees through the long race of life even when the anatomy is 100% modern and normal. Trying to get to the finish line on a knee that used 5-million-year-old blueprints is usually going to involve some limping.

IT CAN ALWAYS BE WORSE

I burned by the medial meniscus earlier because it does not vary in primates. From golden lion tamarins and capuchin monkeys to gorillas and humans, all primates have, under normal circumstances, a C-shaped medial meniscus.[16] In rare instances, however, instead of firing up the equivalent of a flip phone (a discoid lateral meniscus), the human knee goes all the way back to a rotary landline. It goes all the way back to a meniscus that predates even primates. This is the case with the discoid medial meniscus.

A few years ago, I had a student, Hailey, who came to class wearing a large knee brace. This wasn't one of those little over-the-counter types you see weekend warriors wear during a city-league game. This was the type of bulky brace people wear after surgery. I chatted with Hailey and learned about the history of her troublesome knee. She had been fighting knee trouble for years and would end up needing additional multiple surgeries in an attempt to live pain-free.

Hailey grew up in a small town in Eastern Oregon. She played several sports and became an accomplished athlete in volleyball, basketball, and softball. As a sophomore in high school, she was well on her way to her dream of playing college basketball when she felt her left knee pop while taking a swing during the state championship softball game. An MRI a few days later revealed a torn medial meniscus. It also revealed that the medial meniscus in her injured knee was discoid shaped. The condition of a discoid medial meniscus is much rarer than a discoid lateral meniscus, occurring in only about one in every 1,000 people.[17]

Following multiple other meniscal tears and surgeries, Hailey eventually had her entire misshapen left discoid medial meniscus removed. During her senior year of high school, she received a transplanted medial meniscus from a cadaver. Hailey and her surgeons hoped the transplant would put an end to the problem. Over the years, however, with little to no medial meniscus to speak of, the rest of Hailey's knee had taken a beating, leaving other anatomical features at risk. Her knee became even more unstable, and in her first year of college it started to hurt again. The following summer, an MRI revealed a torn quadriceps tendon.

After two more knee surgeries to repair the torn quad tendon and clean up some scar tissue on her transplanted meniscus, Hailey now has quad activation in the medial side for the first time in more than two years. She has had eight knee operations before the age of 21 and knows an absurd amount of detail about the anatomy of the knee for an undergraduate student.

An interesting sidebar to this story is that Hailey has an identical twin, Hannah. With the exact same set of DNA, Hannah's knees developed normally. During her fetal development, Hannah's cells successfully ignored the crossed-out sections of the "story" and put together two modern knees. Hannah went on with her lucky knees to play volleyball in college. Hailey's cells, with the same set of instructions, at the same time, in the same womb, got one of the knees right but messed up the other one.

As a final note, Hailey has not let all the knee problems deter her love of sports. She doesn't play as much basketball these days, but she has taken up rowing with the same type of dedication she has always shown. I can only hope she didn't end up with some kind of crazy, atavistic rotator cuff.

THE ACL EPIDEMIC

Sore, torn, and generally problematic menisci are only part of the battle with bipedal knees. The other significant evolutionary baggage of the human knee brings the knee chapter full circle and back to my "little A-frame" and torn ACL. The switch to two feet was not kind to the ligaments of the knee. The first reason for this is the same reason menisci are injury-prone. We put all the pressure on just two joints. Dogs, squirrels, panda bears, gorillas, and all the rest of the terrestrial, nonhopping mammals spread the weight of their bodies over four knee-like joints. I'm not sure if kangaroos and wallabies have chronically sore knees, but I'm sure someone Down Under is looking into it.

With all the weight of humans packed onto two legs, the human knees were bound to suffer. The elbows were the beneficiaries of this deal, and as a consequence, you never hear about total elbow replacements. Elbows do not have the weight of the body resting on them. People end up with sore elbows only by doing things like hitting too many tennis balls or repeatedly trying to throw a baseball 100 mph (161 kph).

The increased weight burden is not the only problem for the human knee ligaments: there is also a posture issue. As I learned with my painful squatting experiment, in a bent-leg or crouched stance, the thigh muscles take on the bulk of the energetic cost of maintaining posture. As a consequence, the muscles quickly fatigue if a human assumes a crouched position. For bipedal life to be possible, humans had to eventually rise out of their crouch and become straight-legged. In the straight-leg arrangement, with the femurs angled inward, the energetic burden of support shifts from the muscles to the cartilage, bones, and connective tissues holding it all together. The ligaments and tendons of the knees bear a significant amount of the burden of a bipedal existence and are particularly injury-prone in humans.

With a quick glance at ACL injury statistics, one thing jumps out straightaway. Females have a much greater relative risk of tearing their ACLs than do males. Ultimately, more males experience tears than females, but only because there are more males engaging in the types of sports in which ACL tears are common. If you control for the number of individuals participating, females are anywhere between two and eight times more likely than males to tear their ACLs.[18] It is helpful in these situations, as in all scientific situations, to control all but a single variable. So focusing on soccer players, for instance, the female-to-male injury ratio is 2.67:1. The ratio for basketball players is 3.5:1. What this means is a woman playing basketball is 3.5 times more likely to tear her ACL than a man playing basketball. Each year, approximately 5% of female college athletes playing year-round basketball or soccer tear their ACLs.[19]

To illustrate the point with a specific team, consider the adversity faced by the 2017/2018 Notre Dame women's basketball team. The team lost four of their most significant contributors to ACL tears in one year. A solid one-third of the team had season-ending knee surgery. The remaining players rallied hard and lost only three games during the regular season. They lost more players to

ACL tears than they lost games! Incredibly, they overcame all the knee-based adversity and won the national championship. Their roster was so depleted they ended up playing only six women in the title game.

The gender difference in ACL tear rates turns out to be a real rat's nest to untangle. There is no universally accepted explanation for why women are more prone to ACL tears than men. One idea bandied around frequently is that the ACL is more prone to tearing in a female because the female knee sits at a different angle relative to the hip compared with the male knee. Some believe the female skeleton has a trade-off to make between the best angle for bipedalism and the best angle for birthing children. It makes for a nice, tight story, but to date, there is not much evidence to support the hypothesis. Whether or not the anatomical differences are significant is itself a debatable point, and ultimately, studies have not found a significant link between the angle at which the female knee sits and its likelihood to tear.[20]

Another, more straightforward, explanation makes the most sense to me. The female ACL is smaller than the male ACL, even when standardized for body weight.[21] There is convincing research demonstrating that ACL tears are more likely to occur, independent of gender, in individuals with smaller ACLs.[22] A thicker ligament may be less apt to snap when placed under strain. Women may be more likely to tear their ACLs simply because many of them have smaller ACLs to begin with.

BUY A BALANCE BOARD

There is one final piece of the ACL injury epidemic story. As mentioned at the beginning of the chapter, most ACL injuries occur in a noncontact manner, such as when landing from a jump or when experiencing a rapid deceleration event, like pivoting or cutting. In the case of landing after jumping, the ligament is more likely to tear if the knee is extended. Females tend to have a more erect landing posture, which may also be part of the explanation behind their higher rates of ACL tears.[23]

Armed with this understanding, some trainers have started injury prevention programs with young athletes in an attempt to train their bodies to land and cut differently. They hope to reduce ACL injury rates through a series of exercises using balance or wobble boards. The exercises increase the strength and flexibility of the joint and also train proprioceptive sense. Proprioception is the awareness of the position of the various parts of the body. By putting patients through controlled jumping and balance exercises, it may be possible to train them to jump and cut in a safer way. Multiple studies from several countries have shown such training reduces the likelihood of ACL injuries in a range of sports from soccer to basketball to team handball.

I'll never know exactly why my knee blew out during that intramural game. Describing the notch my ACL fits into as a "little A-frame" was my surgeon's polite way of saying my ACL was rather dainty. I don't exactly have the legs of a middle linebacker, and maybe there was never any hope for my delicate ACL.

My training may also have been to blame. I probably spent too much time shooting three-pointers and not enough time working on proprioception and proper jumping and cutting techniques. After all, if you blow out a knee and lose the ability to jump, that well-practiced jump shot is pretty much . . . well . . . shot.

5

The Honey Holiday

Which animal, when it really needs to hustle, can switch away from its typical locomotion and run on two feet?

a. cockroach

b. dingo

c. field mouse

d. tarantula

My first job out of graduate school was at a small two-year college in northern Wisconsin. I grew up in Colorado with mountains all over the place, so the topography of the upper Midwest was a shock to my system. I am very much a "when in Rome" kind of person, so even though the land was far flatter than my template, I quickly embraced the recreational activities of the area. Paddling opportunities are abundant in the Northwoods, and I spent many summer days floating down the Red Cedar and Namekagon Rivers fishing for bass and counting turtles lazing around on logs. Winters were harsh and endless, but snow on the ground for months on end made for great cross-country skiing. Once the lakes froze over, it was time for ice fishing, a frostbitten activity best enjoyed, in my experience, with copious amounts of alcohol.

We had lived there just a short time before Isle Royale National Park caught my attention. Isle Royale is a unique sanctuary, being completely free of cars and roads and accessible only via water. Biologists know the Lake Superior island well because it is home to the longest predator–prey study ever conducted. There were already moose on the island when a pair of wolves crossed over to it from Ontario on an ice bridge during a particularly hard winter in 1949. Scientists started studying the population dynamics of the wolves and moose in 1958 and have kept it up since.

I was intrigued by the island and its rich biological history and decided that walking the 40 miles (64 km) from one end to the other would give me a nice feel for the place. After arriving in the afternoon via ferry, I got a few miles into my trek and made camp for the night. The next day I planned to cover about 15 to 20 miles. I was not overly worried about it, having put in long days several times in the past on other trips.

The majority of day number two was enjoyable. I ran into just a couple of other people, the weather was great, and I saw a pileated woodpecker so large it looked like a bedazzled crow. It took six or seven hours to cover the distance I had planned. A few hours into the day and my feet started to hurt. I expected they would, as I had walked more than 15 miles on other occasions, and each time my feet had objected somewhere around the 15-mile mark. I don't usually pay attention to how firm a trail is, but the Isle Royale trails were particularly hard. It felt like I was backpacking on pavement. As I slept in my tent on night number two, it started to rain. My feet were aching, but I was confident they would be fine in the morning.

After I broke camp the next day, I quickly discovered my feet were decidedly not fine. I could barely walk. Even with the rain-softened ground, both of my feet were beat up five minutes into the day. They had completely blown out on me. It turns out the combination of standing while teaching and sitting behind a desk (the two main physical components of my job) had not done much to toughen up my feet. They were covered in blisters and hot spots on their way to becoming blisters. There was no option other than to suck it up, but every step was agony. The day before I had whistled and smiled my way down the trail, and now I grimaced and cursed with every footfall. I could have stretched my food and decreased my daily mileage, but I had a ferry to catch. I needed to keep moving, even if it involved some pain.

It took me all day to cover the 10 or 12 miles I had planned for day three. I was lucky Isle Royale is so far north because, in the summer, it was light enough for me to limp around deep into the evening. I pathetically hobbled around in the waning light, half wishing for a wolf to take me out of my misery. I managed to make my scheduled ferry, but I was not sure my feet would ever recover. I left the island with my tail between my legs. All these years later, the first thing I think about when I reminisce about Isle Royale is neither the scenery nor the wildlife nor the beautiful solitude. I think about the miserable pain of my feet and how relieved I was when I emerged from the woods and didn't have to walk anymore.

AMATEUR TREE CLIMBING

Human feet are problematic because while they may work as shock absorbers now, in our recent arboreal past they were structures used primarily for agility and flexibility. This disconnect between the past and present limits their effectiveness at taking a pounding. Agility and flexibility are great traits for grasping and climbing, and as primates, humans are naturals at tree climbing. If you put any group of small children together in an environment with climbable trees, it will not be long until a few of them have ascended into the branches. Most elementary school playgrounds have jungle gyms and other contraptions so kids can scratch the burning itch to climb.

We were out to dinner one night with some other families at a place with outdoor seating. The kids were all playing in a little area off to the side of the tables bordered by a retaining wall. The wall went straight up and was at least 20 feet tall. It was made out of cinder blocks, with little indentations between each block, perfect for little feet. At one point, I looked over and saw that one of the boys, who was only five or six years old, had started scaling the retaining wall. He was already several feet off the ground before I told him in no uncertain terms the wall was not for climbing. Had I not noticed his scampering, I have no doubt he would have zipped right up the entire wall. The desire to climb is built into the DNA of many humans.

Humans are, however, extremely limited in their tree-climbing skills compared with the other primates. In the few million years since our ancestors descended from trees, the human foot has undergone dramatic change. It has become significantly less flexible. The main responsibility of the foot is no longer gripping and grabbing but absorbing the pounding of walking on the hard earth. The foot does not always do a perfect job of absorbing the pounding (as I discovered on my Isle Royale trip), but it also cannot grip and grasp the way it could have in the past. As with so many other human features, the foot is mired in a transition between very old and relatively new demands.

PROFESSIONAL TREE CLIMBING

Sometimes when you go to climb a tree, no low branch is available. For me, not having low branches brings any thoughts of tree climbing to an immediate dead end. Even if the trunk is the right size (skinny enough to get my arms around but thick enough to support me), I cannot wrap my arms around a tree, wedge my feet up against the trunk and scurry upward. My ankles are not flexible enough for such acrobatics. I need branches. But there are a few groups of humans who are totally undeterred by the prospect of shimmying up a naked tree trunk.

The Twa are a group of hunter-gatherers who live in Uganda,

and they do not mess around when it comes to tree climbing. They are able to scramble up trees because their ankles flex to an unbelievable degree. The upward motion of bending the toes toward the shin (or the fingers toward the wrist) is known as dorsiflexion. Standard levels of dorsiflexion for the ankle are in the range of 15 to 20 degrees. I'm not the most flexible chap and I don't think I can dorsiflex my ankle past about 10 degrees. My angle of dorsiflexion certainly does not exceed 20 degrees. The tree-climbing Twa can dorsiflex their ankles an astounding 45 degrees.[1]

Why are the Twa so interested in, and adept at, climbing trees, you ask? Honey. Data from other African Pygmy populations have shown that hunter-gatherer groups acquire the bulk of their calories from honey during the honey season. They consume as much as 0.83 kilograms (1.83 pounds) of honey per person per day during the three-month period sometimes referred to as the honey holiday.[2] That's a lot of honey. My daughter would have trouble as a member of the Twa because being sticky is currently her worst nightmare. I would imagine it is difficult to avoid being sticky when eating bucketloads of honey.

The bees' nests containing the honey are not close to the ground. They are found more at a don't-try-this-at-home kind of height. In one study of honey collectors in the Democratic Republic of the Congo, the average nest was 19.1 meters (63 feet) up in the air, with some as high as 51.8 meters (170 feet).[3] Some Twa climbers now use modern climbing equipment, but others still free-climb with nothing more than a vine wrapped around their waist and the trunk of the tree. Falls from heights above 19.2 meters are fatal 100% of the time, and the odds are not terribly favorable at lower heights. We know this because of a research article in the journal *Forensic Science International* with the no-nonsense title "Risk of Dying after a Free Fall from Height."[4] So, first the honey gatherers risk life and limb to climb up to the nest, and then, once they reach their destination, they negotiate with a hive of bees to collect the reward. And don't forget, they still must safely make their way back to the ground, honey in hand. The whole ordeal has gotta be awfully sticky.

SHIFTY HAMMERS

The human foot is a square peg trying to fit in a behavioral round hole. It is made up of a whole gob of bones that were useful when flexibility was at a premium, but now all those bones cause an array of problems when asked to work as shock absorbers. Something built to take a pounding works best if it contains a minimum number of parts. The best example is probably a hammer, which has exactly two components. Each human foot has 26 bones. A hammer made up of 26 parts would undoubtedly come apart upon impact, rendering it useless.

Jeremy DeSilva, professor of paleoanthropology at Dartmouth University, is one of the leading researchers in the field of human evolution. He specializes in early human foot and ankle evolution and has made the observation that when engineers make a prosthetic foot, it is constructed with many fewer pieces than a natural foot.

DeSilva's observation about prosthetic feet stands in stark

contrast to the construction of prosthetic hands. Rather than trying to make a rigid device made of few components, engineers who make prosthetics for upper-extremity amputees have the challenge of recreating a highly dexterous appendage. Hands are a nice example where, instead of failing us, our anatomy truly shines. Employing the remarkable opposable thumb, we still use our hands like the primates we are. Even adults who never climb trees and avoid monkey bars use their hands every day in ways that require them to be agile and nimble. From dusk to dawn we pick up children, peel bananas, grip steering wheels, type on keyboards, chop vegetables for dinner, turn faucets to draw baths, push buttons on TV remotes, and flip the pages of books before turning out the lights. Our incredibly manipulative hands are yet another feature that sets us apart from the wild beasts.

But the poor foot, with a similar set of many interconnected bones, ligaments, tendons, and muscles, took on a very different job when asked to handle significant amounts of bipedal walking. As a result, our feet twist, slip, and sprain and leave us with a whole raft of foot and ankle issues that take years of training to learn how to properly treat.

CLIMBING THE MEDICAL TREE

Everyone knows that doctors train for a long time before plying their trade. In some countries, like the United States and Canada, future physicians finish a bachelor's degree while completing a regimen of prerequisite courses, take an entrance exam, and then apply for admission to medical school. In those countries medical school lasts four years. Students in Europe and Australia can apply to medical school directly after high school, but then most programs last six to seven years. Under both models, new doctors complete a residency that lasts another three to seven years, and there may be some additional training in the form of fellowships or further specializations. Everyone from general practitioners, to pediatricians, to ER docs, to orthopedic surgeons, to OB-GYNs, to internists, to psychiatrists, to endocrinologists, and just about everything in between follows a path that takes a decade or more to complete. After racking up a gazillion dollars in debt and pushing on into their 30s, freshly minted docs are ready to go out and save the world.

There are some interesting exceptions. The routes just described are not the ones taken by those who study teeth. Because teeth are so painfully problematic for so many humans, there is an entire other medical umbrella dedicated to their care. The vast majority of anatomical features do not get their own set of schools. Aspiring doctors cannot go straight into a professional medical degree that focuses exclusively on the elbow or the wrist. There are elbow and wrist specialists, but they become those specialists after first going to the same medical schools the heart, liver, and uterus people go to. After medical school, the future elbow and wrist docs go off and do a residency in orthopedics.

The system evolved this way because it is much rarer to have to visit an orthopedic specialist than it is to visit a dentist. I've experienced exactly one event that necessitated seeing an orthopedist—the time I blew out my knee. On the other hand, with a few extra visits sprinkled in for fillings or pulled teeth, on top of routine visits for cleaning, I am probably pushing 100 total trips to the dentist. There should be dental punch cards like there are at coffee shops. Fill up the card and you get something like a giant stuffed tooth or one free filling or one visit where they promise not to tell you to floss more often. Teeth are one of those parts of the body that do not work particularly well without constant maintenance, supervision, and professional help.

There are a handful of other exceptions. The study and treatment of eyes is another obvious one. Without special schools for optometry, the world could never crank out enough eye specialists to deal with all our nearsightedness and astigmatisms. The point is, the supply matches the demand. Any anatomical area that needs its own entire branch on the medical tree clearly troubles a great number of people, as we saw with all the problems covered in the first section of the book.

Feet are another feature with their own set of postbaccalaureate schools. The regular medical school path cannot crank out enough specialists to treat all the people with foot problems. Sure, there are foot and ankle orthopedic surgeons who went to med school, but the supply of foot orthopedists cannot keep up with the demands of the limping populace. Thus, there is a whole separate set of schools for students to become podiatrists.

Future doctors of podiatric medicine go straight into studying the foot after finishing their undergraduate degrees. They don't dabble in a wide variety of disciplines early in their careers like other future doctors. After four years of classes and training at an accredited college of podiatric medicine, they then complete a two-year podiatric residency, and for many of them, the residency extends out to three or even four years. Some foot doctors spend up to eight years training on the foot and lower leg before going into practice. By comparison, eight years is more time than neurosurgeons spend in their residencies, and neurosurgery is the longest residency there is. By this one calculus, working on the foot is more complicated than working on the brain . . . but maybe don't mention this to your neurologist or neurosurgeon.

WELCOME TO BUNION TOWN, PLEASE TAKE OFF YOUR SHOES

Compared with the relatively minor tweaks experienced by something like the knee, the foot needed a complete makeover to handle the challenges of bipedalism. Going from a structure used to grip and grasp to one used for walking long distances is more than a simple remodel.

The most obvious change occurred with the position of the big toe. Human big toes are in line with the rest of our digits, compared with the other great apes, who have opposable big toes that, for all intents and purposes, act like opposable thumbs. The abducted big toes of the other great apes cause them to more closely resemble human hands than human feet.

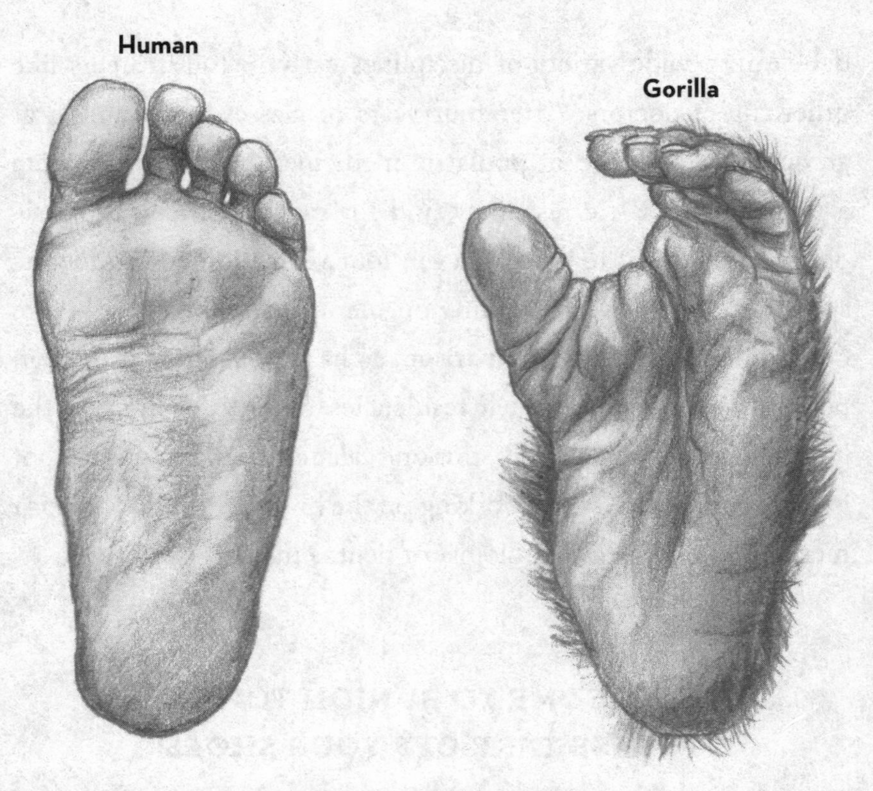

Big toes in humans are lined up with the other digits. All other great apes have opposable big toes that are better at grasping (like in the gorilla foot shown here).

Understanding exactly when and how such a change took place is no simple undertaking. Nothing comes easy with paleoanthropology. The foot specialists have always had it particularly hard because of a historical dearth of decent hominin foot fossils. It seems like there are always plenty of skulls and vertebrae to go around, but most of the unearthed fossilized hominins look like they had their feet chopped off. It's almost as if something came along and gobbled up the feet before sediment covered the rest of the body.

Most of the foot bones discovered in the 20th century were isolated examples. Those bits and pieces never allowed for an understanding of the complete picture of the evolution of the human foot. Even the holy grail of hominin fossils, Lucy, had only three of the possible 52 foot bones with her skeleton.

The lack of decent fossilized foot specimens has required paleoanthropologists to work on occasion with the impressions made by feet instead of the feet themselves. The incredible footprints preserved at the Laetoli site in Tanzania show that by 3.6 million years ago, the big toe was getting in line with the rest of the foot. The footprints unequivocally show the impression of the big toe next to the other digits, allowing for the push-off in the final part of the hominin gait, which is a quintessential feature of modern bipedalism. After generations of trying to piece the story of the foot together with scant hard evidence, the tide turned with the discovery of fossils like the aforementioned species *Ardipithecus ramidus*, or Ardi. In Ardi, whom we first met in the chapter about knees, the paleoanthropologists finally found an example of an early hominin foot. Ardi lived 4.4 million years ago, more than a million years before Lucy, and clearly had an opposable big toe. Based on other features in the foot, scientists believe Ardi was bipedal, but the big toe had not moved around yet to line up with the other digits.[5]

After countless generations with room to breathe away from the other toes, the big toe does not always play nice now that it is all jammed in there next to the other digits. It can push up against the next toe and create some extremely uncomfortable pressure and put the sufferer firmly on the path to Bunion Town. A bunion is a painful, bony bump on the inside of the foot at the joint of the base of the big toe. It is not lonely in Bunion Town. Nearly 25% of the global population between 18 and 65 years of age experience bunions, with the numbers bulging upward to 35% for seniors.[6]

Of course, as with any story, there are two sides to the bunion tale. It is not fair to place all the blame on our ancestors and their opposable big toes. Lest the pendulum start to swing too drastically to the nature side of the nature-versus-nurture spectrum, we should give some attention to nurture for a moment. As with the bulk of the issues covered in this book, and frankly biology in general, the complete explanations typically lie somewhere in the middle of the nature–nurture panorama and not out at one of the extremes.

In this case, feet are bunion susceptible, and then we force the issue with narrow shoes. Next time you are around a baby, check out their feet. More often than not, you will see their toes have a nice wide spread. It's not long, however, before kids' feet are forced into narrow shoes. It seems that somewhere along the line we got the shape of shoes all wrong. Call me crazy, but shouldn't shoes be shaped like . . . feet? Shortly into life, people cram their feet with nicely spaced toes into shoes that are wide at the balls of the feet and narrow in the toe boxes. This works out fine for people whose feet are destined to naturally narrow during their development. It does not work out as well for people whose feet are meant to stay wide, with more breathing room between their toes.

Part of the problem in the United States is that very few people even know the width of their feet. Shoes sold in many other countries tend to give measurements for both length and width, but the whole US shoe industry seems much more interested and focused on length than it does on width. Most people in the United States can quickly rattle off the length of their feet but do not know the width. If they do, it is a sure sign they have had some foot trouble. (I know it has something to do with letters, but I have no idea what my measurement is.) We are totally accustomed to feeling for space between our big toe and the end of the shoe, but what about the space from side to side?

All shoes should have a system like pants, with standard measurements for both length and width. Of course, then it would be just as much of a pain to buy shoes as it is to buy pants, and nobody wants to live in that world. I do appreciate being able to walk into Costco and grab any size 11s and know they will more or less fit, even if they cram my toes together. I suppose I will not be as glib about it if I ever develop bunions.

With a little focus on nurture, do not swing too far to the other end of the pendulum and think modern society causes all foot and anatomical problems. The famous anthropologist Margaret Mead supposedly found individuals with bunions in island populations in the South Pacific where people had never worn a shoe a day in their lives. At the same time, swinging back the other way, the problem is more common in women, and many people believe women are more bunion-prone because they are more likely to force their feet into narrow shoes and high heels. It is not a simple topic. You can bounce back and forth all day between nature and nurture. It is a multifactorial issue with evolutionary, hereditary, and cultural influences.

THE FOOT BONE'S CONNECTED TO THE HEEL BONE

Speaking of buying shoes at Costco, my *God* that place has hard floors. Just a little bit of time spent trudging through the aisles looking at all the six-gallon jars of mayonnaise and I can start to feel it in my feet. I can't even imagine working there. Costco employees must have to spend 10% of their paychecks on insoles. I'm sure they can get a Kirkland 12-pack of them at a fabulous discount. The repetitive striking of feet on a solid surface brings up one of the biggest problems with the foot: the simple weight distribution issue. We are not small mammals, and with only two feet instead of four taking all the day-to-day pounding, the wear and tear of walking around is not a trivial topic. If the surface is hard enough, like the floor at a big box store, the simple act of standing all day can lead to a significant amount of pain.

The heel takes the brunt of the impact of each step, and humans have very large heel bones compared with the other great apes. Analysis of both fossilized footprints and the limited fossilized calcanei (heel bones) suggests bulky heels have been around since shortly after the advent of bipedalism. Australopithecines walking 3 million years ago were doing so with large heels and a prominent heel strike as part of their gait. Hominins with thicker, tougher calcanei would have been able to wander farther from camp. They could have walked longer distances on consecutive days and brought back more food. Slowly but surely, generation by generation, species that had become very comfortable in the trees became more and more comfortable on the ground.

OFF AND RUNNING, SLOWLY

With features like an inline big toe and a nice, fat heel, by 3 to 4 million years ago early hominins were becoming adept bipedal walkers. Walking, however, was not going to cut it forever. Our ancestors needed to be able to move faster. The trigger for the next significant leap forward was, again, climate change. The African environment began passing through extreme cycles of aridity starting around 2.8 million years ago.[7] The wet forests blanketing East Africa to that point became fragmented. Dinner would have become harder and harder for hominins to acquire as habitats became chopped up by the fragmentation. With fractured habitats, it would have been necessary to travel longer distances to forage. At this point, having dabbled in bipedalism for a couple of million years, hominins were ready to take off the training wheels. This is the moment in time when the first members of our genus, *Homo*, appeared, and one critical aspect separating them from earlier ancestors was their ability to travel long distances on foot.

As you might recall from the earlier discussion of teeth, the hominin body was undergoing another major change at that same point in time: the brain had begun to increase in size. Meat became a highly prized commodity for our ancestors. They needed a calorically rich resource to feed their growing brains and also to allow them to tolerate the new, longer distances required for foraging that were the result of the changing environment. The math does not work out in your favor if you walk all day and end up with only a couple of nuts to show for the effort. What you really need at the end of a long day on your feet is a nice, fat, juicy steak.

Tracking down a juicy steak is where running comes in. Daniel Lieberman (the Harvard paleoanthropologist whose research came up in the first chapter—remember the hyraxes on the hard and soft food diets?) and his colleagues suggest running may have evolved in the context of hominins gathering meat.[8] This clearly was not accomplished via sprinting, because humans are rather slow out of the blocks over short distances. If the Olympics were a multispecies event, humans would definitely not make the podium in the mammalian 100-meter dash. We would not even get out of the qualifying heats. We might not even survive if we drew a carnivore in the lane next to us. It would be awfully exciting to watch, though. Someone should make the Mammalian Olympics a thing. Everyone would want to see the world's fastest man running in lane four between a tiger and a gazelle.

Two of the main reasons humans would get trounced in Mammalian Olympic sprint events are the structure of our feet and the nature of our gait. Our feet are not built for sprinting speed. Humans use a primitive style of terrestrial mammalian movement called plantigrade locomotion, in which the bulk of the foot contacts the surface of the ground. When humans walk, the heel hits the ground first, in a process referred to as rear-foot strike by the biomechanics researchers and engineers who study locomotion. Landing on the heel makes for a very economical walking gait. In part because of our rear-foot strike, humans are able to cover long distances very efficiently when walking as opposed to running.[9]

The running story is more complicated. Lieberman and his colleagues discovered that how you run depends on your upbringing. People who grew up in shoes typically run with a rear-foot strike regardless of whether they are running in shoes or are barefoot. Landing on the heel while running causes significant collision forces. Over time those forces can damage the foot and cause pain. In order to mitigate the effects of rear-foot striking, shoe companies make shoes with fat heels to minimize impact. Cushy heels sound well and good, except those raised heels effectively force runners into a rear-foot strike. Fat-heeled shoes also deaden the proprioceptive sense of the body, allowing runners to run in ways that may eventually lead to injury, even though they cannot sense the problem while running because of their fancy, squishy shoes.

In their study, the biomechanics researchers included runners from the Rift Valley Province in Kenya who had never worn shoes. In their article in *Nature*, they discuss how runners who have never been shod naturally land on their forefeet instead of their heels, with the result that "barefoot runners who fore-foot strike generate smaller collision forces than shod rear-foot strikers."[10] They still make contact with their heels, but by running in a toe-heel-toe manner, they minimize the impact on their feet.

This research caused the running world to flip out and spawned the barefoot running revolution that began in earnest a few years ago. There are entire books written about the subject, and all the major shoe companies invested considerable time and energy in marketing minimalist shoes. Obviously, they cannot have everyone running around barefoot or they would have nothing to sell. The mucky-mucks at shoe companies probably love the idea of minimalist shoes since they can still sell insanely overpriced shoes using even fewer raw materials in them. The obvious question is, how much can barefoot running help people who have been in shoes their entire lives? It remains an open question, and scientists should be able to find plenty of subjects to study as there is now a large cadre of people running around barefoot or in minimalist shoes.

I will not be one of their guinea pigs. A few years ago, I got caught up in the hype and tried fitting my feet into toe shoes. My toes are pretty crammed together, and after a few minutes of trying to wrestle each toe into the appropriate slot, I gave up. If I ever become part of the revolution, I will have to fully commit and opt for barefoot running. For now, I do not have any real problems with my feet (unless I try to walk 20 miles in a day), and I say if it ain't broke, don't fix it.

MARCH OF THE PLANTIGRADES

There are quite a number of other mammals that, like humans, use plantigrade locomotion, including bears, rodents, weasels, and raccoons. Plantigrade locomotion is a very stable way of getting around (I cannot think of anything more stable than a bear), but it also cuts down on speed. Don't get me wrong: bears are not slow, they are just not as fast as cheetahs and antelopes. A plantigrade foot has many bones down at the end of the leg, and all those bones drag the leg down, kind of like a heavy ball at the end of a chain. All the contact between those bones and the ground also make it such that the plantigrade animals do not have the natural spring of animals with other gaits. The fence of a bear's enclosure had better be thick, but it does not need to be tremendously tall.

None of the plantigrade mammals are champion sprinters.

Many of them have whole other defense systems necessitated by their comparatively slow natures. Bears are big and tough—problem solved. Some plantigrade animals, like porcupines, have ridiculous prickly adornments. Other plantigrades, like skunks, deter predators by stinking to high heaven. Plantigrade animals like wolverines get away with their lack of speed by being total recluses and also by being really frickin' mean. Some animals with plantigrade locomotion, like snowshoe hares, are, in fact, incredibly fast, but it's because they cheat and switch away from plantigrade locomotion while running.[11]

The somewhat odd group of plantigrade mammals are the primates. Humans and our thumbed friends get around in a variety of plantigrade ways, with some, like the great apes, going so far as to having both the heel and the sole of the foot strike the ground. Other primates have adopted hybrid forms of locomotion somewhere between plantigrade and the other styles in which animals are more up on their toes. Primates do take great advantage of the flexibility that all those different foot and hand bones provide. Primates have around 50% of their bones just in their hands and feet, and all those moving parts allow us to tie our shoes, peel oranges, and fling feces at each other, if you're into that kind of thing.*

---

*As an aside, I know at least *some* little humans are into that kind of thing. I have a friend with a couple of boys who shared a bedroom when they were growing up. When the little brother was a toddler and not yet potty trained, he would wake up in the night, reach into his diaper, and fling the contents far and wide. I've always thought that was a pretty baller move, especially with your older brother in the room. The mom solved the problem by putting his pajamas on backwards, which kept him from being able to undo them and get to the ammo. That's a life hack they do not teach in parenting classes.

Getting back to feet, humans get themselves into trouble because we push the locomotory limits of a plantigrade foot. We ask a foot built for stability and modified for flexibility to take a pounding by traveling over long distances. You will note that there are no great primate migrations around the globe; the primate foot does not specialize in walking across continents like the foot of a wildebeest or zebra. We also exacerbate the issue by getting around on only two feet. If we had stuck with the grooming and poop-flinging like the other primates, our feet would be a whole lot happier.

BALLERINA GIRAFFES

Nothing ever stays the same (except cockroaches and sharks), and mammals became way too diverse a group to stick with only one style of locomotion. In digitigrade locomotion, only the toes contact the ground. This is how cats and dogs roll. The heavier bones are farther away from the ground, losing the ball-and-chain effect of plantigrades. Instead of contacting the ground (like in a human), the metatarsals (the bones between the toes and the ankle) act like a giant spring. It is quite a different challenge to construct an enclosure for a mountain lion than it is for a bear, because pumas can jump 4.5 meters (15 feet) up in the air and leap 12.2 meters (40 feet) while running. Digitigrade locomotion makes for a lot of natural spring and speed.

Human sprinters do their best to mimic digitigrades by wearing special shoes to place them up onto their toes while running. The situation is tenable only over short distances, as it causes a lot of strain on the calf muscles. I learned this the hard way when, as a distance runner for one painful year in high school, I unknowingly bought sprint shoes and spent an entire track season attempting to run long distances in shoes made for running 100 or 200 meters. I did not realize until after the final meet that I had been running kilometer after kilometer in sprint shoes. I just thought my sore calves needed to get stronger.

Then there are the hoofed mammals, the ungulates, with their crazy unguligrade locomotion. They go one step further and contact the ground with giant toenails we call hooves. Most of the bones we think of as foot bones in humans are part of the lower leg of ungulates. I remember a time when the unusual nature of ungulate legs sparked an animated discussion with a biologist friend of mine. We were at a touristy place with carriage-ride horses and debating the position of the horse's "knee" (meaning the intersection of the femur with the tibia and fibula in the rear leg). What looks like the "knee" of a horse is actually the "ankle." The knee is much higher up than most people realize. To complicate and confuse the situation even more, people who work with horses use the word "knee" to describe what is the equivalent of the horse's wrist.

Having most of the bones stuffed up in the leg makes the foot fantastic at taking a beating. The unguligrades are the champions of long-distance walking. From zebras to wildebeests to antelopes, animals walking on their tippy-toes carry out most of the classic, terrestrial great animal migrations. A sneaky, fast digitigrade chases them from time to time, but if they avoid being lion bait they keep plodding along on their feet, which are perfect for walking all day.

This means ungulates are walking around like ballet dancers

en pointe their entire lives. My daughter wanted a ballerina-giraffe-themed birthday party a couple of years ago. It seemed to me like an odd combination at the time, but obviously she was celebrating, in her own way, the tremendous spectacle of unguligrade locomotion in giraffes. Others must feel the same way, as we were able to find a surprisingly diverse array of ballerina-giraffe party supplies. The birthday theme last year was rainbow unicorns, which I can only assume, if they existed, would also walk on the tips of their toes like horses. This year she's pushing for a mermaid theme. Again, not a real thing, but as seagoing mammals, if mermaids were real they would have probably evolved from the same ancestors as whales and dolphins. Whale and dolphin ancestors were, you guessed it, ungulates. I guess the girl just has a thing for cryptozoological animals descended from ungulates.

HUMANS ON THE PODIUM

Despite the many shortcomings of the human foot, if you switched the Mammalian Olympics event to a marathon, we would have a fighting chance to take home a medal. Lieberman's research suggests that around 2 million years ago, hominins became adept at endurance running, possibly as a means of either gathering carrion or literally running prey items to death. Over long distances, a fit human can even keep up with endurance-running specialists like dogs and horses. Adaptations like our abundant sweat glands, lack of hair, and ability to breathe heavily while running allow humans to chase and eventually kill large prey with a method deemed persistence hunting.

We know this was a possibility for early hominins because there are still groups of African hunters who engage in this type of hunting. A large prey animal like an antelope can zip away from a human in a hurry, but it is unable to pant and cool down while galloping. Unable to cool itself through sweating, eventually an antelope stops and pants to control its body temperature. After falling behind in the early stages of the chase, an experienced human hunter can track an antelope and keep plodding along until they catch up again while the antelope is recovering. Once the hunter catches up, the antelope (not quite fully rested) sprints away and the cycle begins again. At the end of the day, after several cycles and a chase of up to 32 kilometers (20 miles), the antelope collapses from heat exhaustion, and the hunter can use a simple tool like a spear to finish the job. Again, this is not just theory. There are groups of modern-day humans who still hunt this way.

The capacity to chase an antelope nearly to death required some changes to the human body beyond those necessary for walking. Long legs make running more efficient, and the legs of hominins like *Homo erectus* (living around 1.8 million years ago) were considerably longer than earlier hominin legs. It works best for running not to carry a lot of weight on the ends of those long legs. As efficient long-distance running became a highly selected human trait, hominin feet became more compact, with shorter toes.

Shoe companies fight this balance today in wanting to make a running shoe with adequate support and shock absorption at the bare minimum weight. In recent years, Nike became obsessed with sponsoring the first athlete to break the two-hour marathon barrier. With the backing of the Nike empire (including a phalanx of pacesetters and every manner of futuristic material built into the shoes), their Breaking2 project was successful in October 2019 when Kenyan runner (and 2016 Olympic marathon champion) Eliud Kipchoge completed 26.2 miles in 1:59:40.* For the record, Kipchoge runs in shoes now, but he grew up running barefoot.

---

*The time of 1:59:40 does not stand as an official world record. The race was set up privately on a flat course with few turns, and Kipchoge used multiple sets of pacesetters. The shoes he ran in have also fallen under controversy as some have argued that the shoe's design provides an unfair mechanical advantage. It is only a matter of time, however, before someone breaks the two-hour barrier in an official marathon. For now, Kipchoge will have to be satisfied with holding the official record of 2:01:39, run at the Berlin Marathon in 2018.

PUT A SPRING IN YOUR STEP

There were no $200 running shoes 2 million years ago, so the body had to supply the spring and support all on its own. The spring came from a variety of anatomical modifications. Just like the eye had to be retrofitted to work on land, the hominin foot had to be retrofitted to allow for efficient long-distance travel. One of the most important modifications was the elongation of the Achilles tendon. Having springy tendons is not a critical adaptation for walking, but they pay big dividends for running. In humans, the Achilles and other elongated tendons in the leg are incredibly efficient at storing and then releasing energy during the stride. It is impossible to know precisely what the Achilles looked like in early hominins because, being soft tissue, it never fossilized. The best you can do is look at the groove into which the Achilles fits in the heel bone, or calcaneus. The calcaneal groove in 3- to 4-million-year-old Australopithecine fossils resembles the groove of a chimpanzee and is markedly different from the wider groove of later hominins and modern humans.

The other important advancement that helped put a spring in our step was the evolution of the arch. During bipedal locomotion, the arch is helpful for maintaining rigidity and mitigating the effects of the force and impact absorbed by walking on two feet. In this sense, the arch helps the foot act as a shock absorber. At the same time, the arch acts like a spring, compressing and recoiling to return some of the energy built up during the first part of a stride.

The basic idea of an arch is not unique to humans. The direction of our arches is what sets us apart. In all other primates, there is a transverse arch, running the width of the foot from side to side. With only a transverse arch, all nonhuman primates are effectively flat-footed. Human feet are unique because, in addition to a transverse arch, we also have two longitudinal arches. The lateral longitudinal arch and the medial longitudinal arch run lengthwise along the foot.

There is considerable arch variation in modern humans. The human arch runs the gamut from people with no arches to people with high arches. Even if not personally afflicted, it seems like everyone has a story about an uncle who could not serve in the military because of a lack of arches, commonly referred to as flat feet. The reports of flat-foot incidence are highly variable. On the conservative end, somewhere around 5% of the population has flat feet. Other reports suggest the condition is much more common, afflicting up to 30% of the population.

One hundred percent of people start off life with flat feet. Most people's arches develop early in life, with the majority of kids showing a typical arch by the age of six or seven years. There are two routes to flat feet. In some cases, the arch never develops, and the individual is flat-footed for life. In other cases, the arch develops normally and maintains its shape for a period of time, but at some point, the tendons and ligaments that make up the arch weaken and the arch falls.

Some people with flat feet are entirely asymptomatic, and they can march around all day just as comfortably as someone with lovely, perfect arches. Other flat-footed people suffer from terrible foot pain and spend incredible amounts of time and energy coming up with makeshift solutions for their lack of arches in an attempt to live pain-free. Figuring out a solution for a painful foot is not simple. Several ligaments make contributions to the arch, and many additional foot and lower leg muscles support it. Pinpointing where the problem is, and what to do about it, is a significant challenge.

Although the particular focus here is on flat feet (because our ancestors had flat feet and tying the anatomy of our ancestors to our modern issues is kind of the whole point of the book), plenty of people with perfectly shaped arches still develop foot pain. Some of those people bring the problems on themselves by training for marathons and running on pavement for 100 miles a week. Other people are stuck on their feet all day at work and subject their ligaments to prolonged stretching and inflammation just trying to earn a buck. The quality of life of many a box-store employee would greatly improve if all those hard floors were made out of gym mats. Lastly, there's a third category of people with foot pain who are simply unlucky. Some people were just destined to have painful feet wholly independent of their behavior.

The most famous of those troublesome foot ligaments is the plantar aponeurosis, also known as the plantar fascia. The other ligaments and tendons can cause trouble, but there is a reason why a Google search of "plantar fascia" pulls up more than four million results while a search of "calcaneonavicular ligament" yields fewer than 50,000. The plantar fascia is a thick band of connective tissue running from the heel to the toes. If you feel the bottom of your foot while bending your toes upward it is easy to feel it stretch. When the plantar fascia becomes excessively stretched it can become inflamed. This can occur because of a fallen arch or flat feet, and it can also happen from repetitive stress.

Inflammation and pain in any of the ligaments of the arch send patients to their doctors or podiatrists on a quest for shoes, insoles, and orthotics to ameliorate their symptoms. Some foot-pain sufferers become desperate enough to undergo surgery. There are many varieties of surgical foot procedures. In some of them, bones are cut, moved, wedged, or remodeled. Sometimes thickened or damaged tendons are entirely removed and replaced with other tendons donated by the patient (in another rob-Peter-to-pay-Paul scenario, like with an ACL repair).

In the case of plantar fasciitis, the question becomes finding the right amount of tension. For the ligament to do its job of supporting the foot and the arch, it has to retain some tension, so surgeons must decide how much of the tension to release.[12] Total release sounds like a great idea for the pain, as 100% of the tension in the ligament is removed. Without a plantar fascia, however, the other ligaments in the foot have to pick up the slack, and if they're not up to the job the arch may collapse (if it hadn't already collapsed). The goal is to remove enough of the fascia to provide pain relief but not so much that the patient ends up with other wonky foot biomechanics. It is a delicate give-and-take with potentially life-altering results for patients. Finding the right balance between excessive and insufficient release of the plantar fascia is one of the many reasons why podiatrists have to go to school for so many years.

THE DAWN OF THE ARCH

It has taken the work of some very crafty scientists to figure out when the human arch evolved. The paucity of foot fossils obviously makes it a challenging subject to study. But even if there were drawers and drawers of perfectly preserved early hominin feet, you run into the old soft-tissue issue once again: ligaments do not fossilize. The bones of the foot outline or border the arch, but the structure itself is ligamentous.

Jeremy DeSilva and his foot paleoanthropology colleagues have figured out clever ways around those problems and placed many pieces in the foot-evolution puzzle over the years. Rather than lamenting the lack of fossils, they have committed to working with what is available. What is available is a modern human population with highly variable feet and plenty of fossilized leg bones from early hominins. They started with the knowledge that tremendous variation exists within the modern human arch. The trick was tying that significant variation to some feature of the lower leg, since the lower leg is typically all you get with Lucy and her fossilized friends.

The larger of the lower leg bones, the tibia, had the variation they were looking for. Specifically, they discovered a correlation between the tilt of the tibia and flat-footedness in modern-day humans. They did this by taking radiographs of 261 living individuals and making a series of detailed measurements. Most people (86%) in the study had an "anteriorly directed set to the ankle."[13] In a handful of people, the tibiae were set perfectly straight at

0 degrees. The most interesting subjects were those with posteriorly tilted tibiae. The 8% of people who fell into this group were more likely to be flat-footed than the individuals in the other categories.

Armed with this new information about modern humans, the researchers looked at the tibial arch angle in the other living great apes (all of whom are flat-footed). The chimps, gorillas, and orangutans all had, on average, posteriorly directed tibial arch angles. This was another piece of evidence suggesting a link between the angle of the tibia and the degree of arch of the foot.

They also analyzed the 12 fossilized early hominin bones they had available to them. Just within Lucy's species, *Australopithecus afarensis*, there were some individuals with negative tibial arch angles (and presumably flat feet) and others with positive tibial arch angles (and presumably feet with arches). Lucy herself had the most posteriorly directed fossil tibia in the study and, as a consequence, was very likely flat-footed.

The results from the fossilized tibia samples show a foot and ankle in flux some 3 to 4 million years ago. A few short million years later and the situation is still fluid. Some people have flat feet, others have well-developed arches, and other people have a foot stuck somewhere between the past and the present. There are even people with unusually high arches, which also lead to problems and foot pain. The human foot is still clearly working through the process of adjusting to a life spent wandering around on the ground.

GIVE THE HAND A HAND

It's important to remember there are amazing benefits that counter all these foot-associated costs related to being bipedal. Let's take a minute to balance the bad with some good. Our feet may suffer from our relatively recent biped transition, but our freed hands shine as a true beacon of evolutionary success.

There are significant differences between human hands and those of our closest living relatives, chimpanzees. Chimpanzee hands are similar to the fossilized hands discovered from ancient hominids, so they work well for comparative studies that try to explain how our hands have changed over the last few million years since becoming absolved of any responsibilities related to transportation.

The most notable difference is the thumb. Chimpanzees and other primates have opposable thumbs, but our thumbs are longer and stronger than in other primates. Combined with our shorter fingers and larger thumb muscles, we have a unique degree of grip strength and flexibility. Scientists note the distinction by referring to the human thumb as fully opposable. We alone can touch the pad of our thumb to the tips of all our other digits. With an opposable but not *fully* opposable thumb, chimpanzees can grip, but their grip of branches for swinging through the trees mostly involves flexing their nonthumb fingers to create a hook.[14]

Freed from the anatomical pressures of an arboreal existence, our hands have changed. With fully opposable thumbs, humans can grip and swing a stick or pick up a rock and throw it with great velocity. These are not things chimpanzees can do. Such behaviors enhance our offensive *and* defensive abilities. A chimp can pick up a stick and flail it around a bit, but without a fully opposable thumb they cannot swing it with any force. A chimp would need to retreat up a tree if engaged in a fight with a mean, toothy carnivore. With my thumbs overlapping my index fingers I can swing a baseball bat as hard as I want without letting go. Give me a decent-sized stick and I can probably stay down on the ground when fighting off a large predator. Before the rabid beast was even upon me, I could also sling rocks at it, which might make it possible to avoid the encounter entirely.

Several million years later and we still occasionally chuck rocks at each other and use our grip for violent means, but now our adroit hands are employed in countless other human ways. The ability to create precise pressure between the thumb and other digits is critically important for tasks like using a knife, weaving, or, in more recent times, gripping a pen or a pencil. The switch to bipedalism spurred changes in our hands that arguably make them as much a part of the human element as our large brains.

SEXY DINOSAURS

We are not as unique in our bipedalism as we might like to think. To knock us down a peg, there are even cockroaches who switch to running around on two legs when they really need to get going. Cockroaches are not a great group for comparative studies, however, since they have six legs and are also so small they can be squashed with a shoe. There are other mammals, like kangaroos, who are obligate bipeds, but they get around by hopping, which is distinctly different from how humans walk.

Birds are the best comparative group for understanding what a bipedal appendage can look like after millions of years of evolution. Like humans, birds have put their freed forelimbs to skillful use. Taking advantage of their modified forelimbs, birds took flight and have colonized every corner of the globe. By sheer numbers, they are the most successful group of terrestrial vertebrates, and it is not even close. The evidence is now overwhelming that birds are the descendants of bipedal dinosaurs who survived the Giant Meteorite of Death some 66 million years ago. Some zoology texts even title the bird chapter "Avian Reptiles," overtly recognizing the true ancestry of birds.

If you are of my generation, you will never forget watching a T-Rex try to snatch a jeep during the great chase scene in the original *Jurassic Park* movie. The notion of quadrupedal lizards transitioning to bipedal lizards isn't such a leap when you realize there are several modern-day species of lizards that can pop a wheelie up onto two legs.[15] The most famous among the group is the common basilisk, or Jesus lizard, who takes it up a notch by running on two legs on water for short stretches. *Tyrannosaurus rex* certainly was neither getting airborne nor walking on water, but there were other bipedal dinosaurs that had feathers and were transitional organisms between dinosaurs and birds.

Velociraptors are the other dinosaurs everyone remembers from

Jurassic Park, screeching like mad as they tried to eviscerate the scientists. Velociraptors were also stuck on the ground, but they had feathers and were members of the group of dinosaurs from which birds eventually evolved.[16] Paleontologists don't know exactly what flightless, feathered dinosaurs were doing with their feathers, which means they need to put on their brown outfits and get back to digging. For the record, my money is on the hypothesis that feathers evolved from scales because they look sexy. Then again, I think sexual selection is always the safest bet as an explanation for evolutionary change. It is certainly a much cooler idea than thermoregulation, which is another, albeit duller, established hypothesis for the evolution of feathers.

Jeremy DeSilva likes to make the comparison between the human foot and the foot of a modern bird that does not leave the ground. An ostrich has just eight bones in its foot, compared with the 26 bones found in the modern human foot. DeSilva notes the ostrich foot "looks a lot like the new design for human foot prosthetics."* With so many fewer pieces, the very stable foot of an ostrich can carry it at speeds in excess of 64 kph (40 mph) without breaking down.

Birds have quite a head start on construction of the bipedal foot and leg. They have been kicking around as bipedal animals for more than 100 million years. The human foot has had much, much less time to work out its problems. It is the dawn of our bipedal-

---

*DeSilva wrote online about the comparison as part of a guest post for the Leakey Foundation on a page called "Why walk on two legs?"

ism, and we're still a little wobbly as the day breaks and we get out of bed. If you are a box-store employee forced to stand all day on a rock-hard surface, try explaining this to your boss. Maybe, just maybe, they'll cut you a break and find you a stool.

6

Baby Got Back . . . Pain

Which rapper praised plump backsides in the first line of his classic song "Baby Got Back"?

a. Ice Cube
b. Ice-T
c. Sir Mix-a-Lot
d. Snoop Dogg

When Julie and I were married, her brother gave us a queen-sized mattress as a wedding gift. We were thrilled because we had been sharing a twin bed handed down to Julie by her grandparents. The little bed had a carved, decorative footboard with holes in it, and sometimes I would wake up with my feet poking through them like they were in the stocks. Sharing a twin bed worked for a certain amount of time, especially when we were broke and the relationship was still young, but eventually, I needed to stretch out my long frame.

When we went to pick out a new mattress, we had to deal with our decidedly different sleeping preferences. I had grown up sleeping on the concrete slabs my parents called beds and had become quite accustomed to a very firm mattress. Julie had always slept on a cushier mattress. Clearly, something was going to have to give.

Admittedly, I have always been the fussier sleeper. For starters, I need it to be pitch black and dead quiet. Julie says I have paper-thin eyelids. For years I've slept with earplugs and an eye mask, like a total diva. It also takes me a while after lying down to fall asleep. In contrast, with zero paraphernalia, Julie is able to fall asleep in about five seconds. I sometimes lie there listening to her breathe, wondering how on earth anyone can fall asleep so quickly.

I am also much better at complaining than Julie. I consider it an important life skill. She will sometimes be sick or in pain for several days before I figure out something is wrong. That could never happen with me. Within five minutes of getting a splinter or a tickle in my throat, everyone within a five-mile radius is aware of my health status.

You can see where this is going—we got the firm mattress. I'm sure Julie decided being uncomfortable was easily the preferred option compared with listening to me grouse and moan about not being able to sleep. She soldiered on through those early years of sleeping on a mattress that was as stiff as a board and never complained unless I worked hard to draw it out of her.

Skip forward almost 10 years and it was time to get a new mattress. We had purchased our first home and had graduated to the stage in life where you set up a real guest room. We decided to put the old mattress in the guest room, and thus we faced the same soft or hard mattress decision we had as newlyweds. It seemed only right Julie should get to pick the second mattress. I had hoped all the years of sleeping on plywood would bring her over to the firm-mattress team, but alas, she still preferred the mushy variety. We came home with some cloud-like mattress with a downy pillow top, and I was now the one forced to suck it up. I like to think I took my lumps well during those soft-mattress years, but you can bet I snuck plenty of bellyaching in there.

Jump ahead almost another decade and we are getting close to present time. For mattress number three we decided to roll the dice and go with the middle option, somewhere between firm and soft. Maybe a mattress that made us both a little uncomfortable—but not as miserable as we could be—was the best bet. Julie, true to form, has complained very little but is not as comfortable as she was on the cloud mattress. I am more miserable than ever. I am waking up in the night in pain and trying new stretching regimens to see if I can solve the problem by getting stronger and more flexible.

This incredibly sound science is complicated by one significant factor: age. It is impossible to hold the variable of age steady when running the personal, decades-long mattress experiment. Kids can sleep on anything. If my daughter were tired, I have no doubt she could fall asleep on a piano bench covered in lava rock. I spent a summer in my early 20s traveling around the United States, working odd jobs with a buddy while living out of my pickup. We slept in the truck bed all summer long and I snoozed like a baby. In my

30s, sleep started to become more uncomfortable, but I still had no problem knocking out for many hours at a time.

It's cliché, but around the time I turned 40, sleep became more difficult. I can already tell I am going to be one of those old men who wake up at 4:00 a.m. and cannot get back to sleep. I'll be up and at 'em with the roosters and have put in a full day before noon. In my younger days, it wasn't unheard of for me to wake up at noon. Julie and I used to be champions at sleeping in. Now, on the rare occasion when we get to sleep in, I can't make it past 8:30 without aches and pains forcing me out of bed.

The answer for us is obviously one of those fancy Sleep Number beds where you can make it into pavement on one side and goose down on the other. One of those has never been in the budget, but hopefully, we'll be able to pull it off for mattress number four in another decade. Otherwise, I'm going to have to figure out exactly what it is old men do at the crack of dawn.

A GLOBAL PROBLEM

The pain driving me out of bed is in my lower back. This whole section of the book (the chapters on the knee, foot, and back) is effectively an exploration of the skeletal consequences of bipedalism, and the biggest bugaboo of walking around on two feet has to be how it has wrecked the human back. I am lucky, because once I'm up and around I'm able to go about my day and be largely pain-free. For so many unlucky others, lower back pain is a debilitating and disabling curse from which there is no relief night or day. In a survey conducted by the American Physical Therapy Association (APTA), 61% of Americans reported experiencing lower back pain. Other reports suggest that, at some point in their lives, up to 80% of adults will suffer from lower back pain.[1]

Lower back pain affects more than just pampered Americans. It is a significant global problem. The British journal *The Lancet* published a study a few years ago about the leading causes of global disability.[2] The Bill and Melinda Gates Foundation funded the study, and it underscored a number of revealing trends.* The system the researchers used for quantifying pain took into account both the prevalence and severity of conditions. In wealthier countries with older populations (e.g., the US, Canada, most of Europe), the leading cause of disability was lower back pain. Other, poorer countries were not free of back pain; lower back pain ranked in the top 10 causes of disability for all 188 countries surveyed. Less-developed countries had the added challenges, on top of back pain, of conditions like chronic respiratory diseases and HIV/AIDS, which are significantly less common in more developed countries. In poorer countries with younger populations (e.g., Mexico, India), depression was the leading cause of disability.

Back pain radiates outward from patients and becomes a societal issue. In the APTA survey, after exercise and sleep, the third category most greatly affected by lower back pain was work. Even

---

*An illuminating figure from *The Lancet* study was published by *The Economist*. Find it in Google Images by searching *The Economist* most common causes of disability.

in countries with smaller economies, the annual cost of lower back pain to society easily adds up to billions of dollars, and it can add up to hundreds of billions of dollars for wealthier countries when considering both the cost of treatment and the loss of worker productivity.

The other issue closely related to back pain pertains to its treatment. Opioid medications like morphine, oxycodone, or hydrocodone are often prescribed for pain management following an acute injury or surgery. In those cases, the goal is to have patients on the powerful drugs only in the short-term to achieve the necessary pain relief. There are relatively few opponents of the use of opioids in this manner to control intense, acute pain.

What is far more controversial and problematic is the use of opioids for the management of chronic pain.[3] Patients in chronic pain are often prescribed these highly addictive drugs and then take them over the course of weeks, months, or years. This type of opioid use requires much more individual discussion and caution on the part of the care provider and the patient. In the worst-case scenario, a patient develops an opioid use disorder and becomes another statistic in the growing opioid epidemic. Surprisingly, there have been very few studies comparing the efficacy of opioid versus non-opioid medications for the treatment of chronic pain. Thus, despite the incredible risks of their long-term use, it is unclear at this point how effective opioids even are in the management of chronic pain. Given that more people are now dying from opioid overdoses in the United States than from breast cancer, the issue clearly deserves some serious research dollars.[4]

CRAWL, TODDLE, CRAWL, TODDLE

Back pain is a universal conundrum for which we do not have a good solution. Conveniently for me, this book is not about solutions. It is about why, as humans, we are uniquely prone to aches and pains. You may have noticed the initial focus of this chapter has been on *lower* back pain. Of course, pain can occur anywhere along the spine. In fact, the aforementioned Gates-funded study reported neck pain as the fourth-ranked cause of global disability after lower back pain, major depressive disorders, and iron-deficiency anemia. Neck pain is really just high back pain, radiating out from the cervical portion of the spinal column instead of the lumbar portion of the spine. The back is going to get you one way or the other, be it high, low, or somewhere in between.

The shape is the issue. At least, the shape is the short story behind the pain of the back. The center of gravity of the body had to shift in order for a quadruped to balance itself on two feet. As mentioned in the foot chapter, a chimpanzee's vertebral column is C shaped, placing the center of gravity well forward of the hips. If chimps force themselves onto two feet, it takes a lot of muscular effort to overcome their natural forward pitch. So chimps typically avoid walking around bipedally. They tend to walk around on all fours, saving the strain on their backs.

Like chimps, human babies are born with C-shaped spines. It makes sense, therefore, that human babies are not born bipedal. As babies get bigger and stronger, their vertebral columns start to develop the other curves that switch their spines from rigid, horizontal beams to more flexible, vertical structures. Early in life, the muscles and curves are developing but are not yet strong enough to hold babies upright. During this period, babies get around by crawling. Once the curves develop and the muscles are strong enough, then they are off and toddling. In the transitional period, you can almost picture kids deciding if it is worth the effort to walk to the other side of the room or if they should just crawl. The first steps can happen as early as six months of age and as late as 18 months, but somewhere around one year is typical.

I was at a party recently where a toddler did something I had never seen before. She had just learned to walk and was in that stage where there is a lot of wobbling and stumbling. At one point she tipped over and instead of going all the way down to crawling, she caught herself on her hands and proceeded to make her way across the entire living room floor walking on all fours. It was fascinating to watch and felt for a second like I had stepped back in time a few million years. Maybe when she grows up, she'll break the record for the fastest 100 meters running on all fours.

TAKE THE CURVES SLOWLY

The lower lumbar curvature shifts the center of gravity back, placing it above and slightly behind the hip joints. This change in the center of gravity allows toddlers to toddle and adult humans to stand all day and not fatigue. Our feet might get sore, but we already dealt with sore feet in the last chapter. The higher-up cervical curvature allows us to balance our large heads on top of our spines without constant contraction of the neck muscles.

Clearly, there are advantages to the change in the shape of the human spine. I, for one, like walking around on two feet. You get a better view of the world, and I enjoy having my hands free at all times. There are, however, some significant cons to go with those pros. The shape of the curves has to be just right or everything goes haywire. Think of the vertebral column as a series of children's blocks. In other mammals, the blocks are lined up horizontally, as if they are laid down on a table. You could not knock them over because they are just sitting there. In a human, the blocks are upright, like a tower. Even before introducing curves, the structure is much more precarious. Now, as one final twist, next time you construct a block tower, try building a few curves into it and see how that goes for you.

The angle and degree of spacing between the vertebrae are critical. If the spacing between the vertebrae is not right, the intervertebral discs can go oozing out of place. This is a very delicate area for anything to be out of place because the spinal cord runs like a highway right through the middle of it all. At each exit on the highway, nerve roots branch off the spinal cord. Pain begins as soon as a disc bulges out of place and contacts a nerve.

The pinching can happen acutely when you bend over to pick something up, or it can happen chronically when the spine does not end up with the right amount of curvature. The inward, or lordotic, lumbar curve needs to be far enough inward to place the position of the spine under the head and to get the center of gravity above the hips. Sometimes there is too much lordotic lumbar curvature. In a confusing bit of terminology, the natural curve is called lordosis and the condition where the curve arches too much is called . . . lordosis (or swayback). Too little lordosis = back pain. Too much lordosis = back pain. The right amount of lordosis = total bliss—at least until you're old, when there will likely be some back pain no matter how perfect your curves are.

Even the more ancestral curve through the middle, or thoracic, portion of the back (the one present at birth responsible for the C-shaped spine of babies) can get out of whack. Kyphosis is an excessive outward curvature, and when taken to the extreme it can cause an individual to appear hunchbacked. As the bones and muscles of the back weaken, the spine can become kyphotic, causing some elderly people to look as if they are on the verge of falling forward.

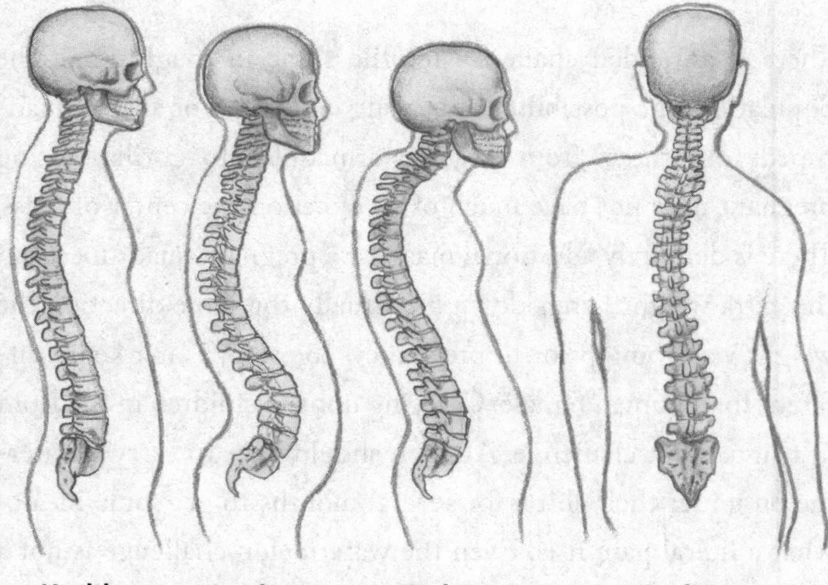

Schoolkids are often evaluated for scoliosis, a different type of abnormal curvature. Lordotic and kyphotic curves are visible only when the spine is viewed from the side. In scoliosis, the spine curves sideways, making it possible to see the condition when viewed from behind. Children are checked during elementary school because it is important to catch the problem early. Many scoliosis cases require only consistent monitoring rather than intervention, but in more serious circumstances, a child may need to wear a brace or even undergo surgery. The condition is considered idiopathic (meaning we do not know what causes it), but it certainly cannot have helped when our ancestors left our stiff, beam-like spine up in the trees and traded it out for a fancy, flexible, curvy model.

LORDOTIC CURVES

There is an added challenge for the spine in roughly half the population: the possibility of bearing children. For all the quadrupeds out there, from dogs to armadillos to gorillas, being pregnant does not have much of an effect on the center of mass. There is definitely additional mass for a pregnant quadruped, but the extra weight hangs down in virtually the same direction the weight was going prior to pregnancy. The story is markedly different for a human female. Carrying unborn children in the front is a significant challenge. All men should have to carry a watermelon under their shirts for several months to get some idea of what a literal pain it is. Even the watermelon challenge is not a fair comparison, because unless you had it surgically implanted, there is no way to get it shoved right up against the bladder. Also, watermelons don't kick.

Being pregnant greatly shifts the center of mass of a woman's body forward. The center of mass needs to have a way of shifting back. Otherwise, as discussed in an article in the journal *Evolution and Human Behavior*, "ancestral women would have been subjected to a nearly 800% increase in hip torque during pregnancy."[5, 6] It's hard to believe it could be worse than it is, but an 800% increase in hip torque would translate to an even greater degree of muscle strain and back pain than is already present during those joyful nine months.

More lumbar lordosis is the key. All humans naturally possess lumbar lordosis, but when women become pregnant, they become uberlordotic. The extreme shift in the shape of the spine during pregnancy pushes the center of gravity back, but it also causes extensive shearing forces (where parts of the body are pulled in opposite directions) throughout the lower back.

Those shearing forces are at the root of the intense and persistent back pain experienced by many women during pregnancy. The best-case scenario is sciatica and a sore lower back. The worst-case scenario is a herniated disc. During the years after hominins became bipedal, there would have been strong selective forces for anatomical varieties that reduced the degree of shearing and pain. Women with bodies more able to cope with a bipedal pregnancy would have been able to forage to a greater degree and provide better care for their growing families.

The shape of men's and women's lumbar vertebrae reveals the strong selective pressure on the back. In both sexes, the penultimate and last lumbar vertebrae are wedge shaped, which helps the spine disperse the pressure placed on the lower back. In females, the lordotic lumbar wedging reaches up to include an additional vertebra. It might not seem like much, but the difference is significant. Having one more wedge-shaped vertebra allows females to mitigate the effects of shearing and stress caused by the extreme degree of lordotic curvature assumed during pregnancy.

OH MY GOD, BECKY

For all humans, and especially for women, there is a fine line between not enough lordosis and too much lordosis. Given the clear benefit to women, in regard to pregnancy, of having the right degree of lordosis, a logical hypothesis is that lumbar curvature in women is a signal of attractiveness and therefore a sexually selected trait. Naturally, some researchers at the University of Texas at Austin gathered up a bunch of dudes from the psychology department to test the idea.

They generated images of women, using Adobe Photoshop, where the only variable among the pictures was the degree of lumbar curvature. There is an incredible amount of natural variation in lumbar lordosis, with an average angle of curvature reported for adult humans of around 30 to 35 degrees, but with a ridiculously wide range running between 14 to 69 degrees.[7, 8] The researchers gave the test subject males stimuli that "captured the naturally occurring range of lumbar curvature in the population."[9] Lo and behold, on average, the male psych students demonstrated a preference. The sweet spot was somewhere in the middle, with an angle of lumbar lordosis a little above the average—somewhere around 45 degrees—getting the UT bros the most worked up.

Because these sex-crazed back researchers are a thorough lot, they took the study one step further. They knew that, in addition to vertebral wedging, buttock mass can also influence lumbar curvature. So they gathered up some more psych students to see how much of an influence buttock size had on male preference. This time, the guys ranked the attractiveness of computer-generated models with the same degree of "buttock protrusion" but with varying degrees of buttock mass and vertebral wedging. In other words, is a tush cute because of the meat in the tush or because it sticks out owing to the curve of the lower back? And the winner was . . . vertebral wedging. Independent of buttock size, men preferred the silhouettes with the sexy 45-degree lordotic curves. So instead of professing his love of big butts, it would have been more accurate for Sir Mix-a-Lot to say, "I like 45-degree lordotic curves and I don't know why." I'm cool giving him a pass there. For one thing, the UT research was still decades away, and it also would have been much harder to rap.

RETRO VERTEBRAE

For some people, back issues arise in one dreadful acute moment. It all starts with a slip off a curb or a brief lapse in judgment lifting something way too heavy. Just the other day, a friend of mine threw out his back stepping on a Lego after putting one of his kids to bed. Other times, the process is more of a slow burn. The pain creeps in and is easy to ignore for the first few years. As the tissues continue to degenerate, the pain builds until it reaches a tipping point.

The curves are only part of the problem. The other major issue is the vertical nature of the human spine. Horizontal building blocks do not have to support weight like stacked vertical building blocks. Bipedalism places a ton of stress on the spine that is simply not present in a quadruped. As a result, humans are uniquely prone to spinal problems compared with other primates.[10]

Recent research has shown that vertebral shape may dictate how effective a person's back is at managing the stress of bipedalism. The data collected support the ancestral shape hypothesis, which suggests some humans may be more prone to back pain, and lower back pain in particular, because they have vertebrae shaped more like ancestral vertebrae than modern vertebrae. Some people have "newer" vertebrae, where the bodies of the vertebrae are heart shaped. In other people, the vertebrae are more retro, with a shovel shape. Naturally, those are two ends of the spectrum, with every variety of possible shape in between.

In one study, researchers divided human lumbar vertebrae into healthy and pathological groups and showed that the pathological lumbar vertebrae had characteristics of chimpanzee vertebrae. Compared with healthy human vertebrae, pathological human and healthy chimp vertebrae had a smaller space for the spinal cord to pass through as well as a shorter connection between the body and spine of the vertebrae.[11] They were also more shovel shaped. It turns out, injury-prone human vertebrae look a lot like normal chimpanzee vertebrae.

What would be helpful is if we could all get images of our spines included in our medical files. I would love to know my angle of lumbar lordosis and if my vertebrae are more human-like or more chimpanzee-like. Mine have been feeling more chimpanzee-like as of late. Knowing trouble is coming would inspire some people to start stretching and strengthening before the pain crops up, likely delaying its onset. In more serious cases, having injury-susceptible spinal characteristics might drive some kids away from activities that place undue strain on the lower back like running, gymnastics, golf, and contact sports and point them toward exercises with less impact and twisting of the core, like swimming and skiing. Granted, you might blow out an ACL while skiing, but if I had to choose between an ACL repair and living with back pain, I would say the lesser of those two evils is the reconstructed knee.

CHIMPANZEE NIGHTMARES

For my peace of mind, a quick sidebar is necessary here to reiterate the relationship between humans and chimpanzees. Chimpanzees keep coming up in studies I reference because they are the closest living relatives of humans, making them a very natural group for comparative studies. The repeated references to chimpanzees make me feel obligated to stress—*humans are not descended from chimpanzees.* Humans and chimps are both descended from a now extinct common ancestor who lived millions of years ago. I wake up in a cold sweat with nightmares about my students leaving college with the human-chimp relationship screwed up in their heads. Now I have the same worry about anyone who reads this book, so I'm glad we got that taken care of. Okay, moving on . . .

OOZED DISCS

Our troublesome intervertebral discs have quite the backstory. The cartilaginous discs have a soft, gel-like central portion, called the nucleus pulposus, surrounded by a tougher outer ring called the annulus fibrosus. Analysis of the cells from the nucleus pulposus has revealed that portions of this softer, squishier section of an intervertebral disc are derived from the notochord.[12]

If you have ever taken a biology class, you have the word *notochord* tucked somewhere within the recesses of your brain. I review the concept in just about every class I teach when we go back over the taxonomic classification of humans. All organisms are classified into a handful of kingdoms (e.g., bacteria, plants, animals), and then those kingdoms are further subdivided into phyla. There are 36 recognized phyla of animals. Most of them you've never heard of, unless you're into super obscure animals like mud dragons, penis worms, and water bears. I mean, I don't pretend to know your life, but I'm betting you haven't spent a lot of time thinking about penis worms.

One important phylum students seem to struggle to remember is our own phylum. Very few people, including those who have learned it before, remember we are members of the phylum Chordata. Even though most of the animals we know and love, including fish, amphibians, reptiles, birds, and all other mammals, are in the phylum Chordata, the term has not caught on like other popular classifications such as mammal, primate, or vertebrate. Even though it is *our* phylum, it proves more challenging for students to recall than other phyla like Mollusca or Arthropoda. Chordata has a serious branding problem.

The defining feature of chordates is the presence of a notochord. The notochord is a beam of tissue that provides a longitudinal structure for the development of a chordate. I will grant you the concept is not as sexy as mammary glands, which give the class Mammalia its name, and maybe this is why nobody remembers we are chordates. The Chordata PR firm has to work with a flexible rod instead of dazzling lumps of tissue that make milk. That's not really a fair fight in terms of what people will remember. Notochords have ancient origins and are present in a few very simple organisms like sea squirts, which I have always found quite humbling to have in my same phylum.

Having taken place hundreds of millions of years ago, it is difficult to know exactly why notochords developed, but some scientists have suggested the notochord provided a rigid structure for the attachment of developing muscles.[13] If you roll the clock back far enough, some of the most primitive ocean organisms could swim only by ferociously beating their cilia and/or flagella, the

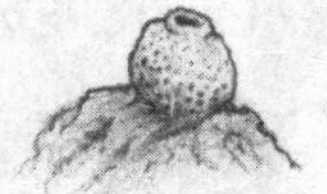

microscopic hair-like projections found on nearly all cells. The development of a stiff beam along the long axis of an animal gave muscles something substantial to push against, which would have improved mobility.

In the vast majority of chordates, the notochord is present only embryologically. Adult animals still need something rigid as an anchor point for muscles, and the vertebral column serves nicely in this capacity. The cells of the notochord transition in most adult animals to provide the cushion between each of the vertebrae. There are some weird exceptions like hagfish, which continue to fly the flag of anachronistic and transitional organisms, where the notochord persists as a rod through adulthood.

Bones are great and all, but they do not work particularly well without some padding between them. The vertebral column is no exception. When the padding wears down, bone grinds on bone and trouble ensues. Intervertebral discs provide some cushion between vertebrae. In humans, they need to do more than just cushion; they have to act as serious shock absorbers. With the advent of bipedalism, the job of the intervertebral discs became far more difficult. In quadrupedal animals, the job of the discs is to keep the vertebrae from rubbing together. They still have this job in humans, but because we walk upright they have to do it with extra weight pushing down on them.

The squishy middle gel of the intervertebral discs can only take so much. It either oozes out during an acute event (watch out for those Legos), or, after getting squashed for decades, it starts to migrate away from its snug little home between the vertebrae. If not confined by the hard outer ring, it can press on nearby nerves and, voilà, you have a herniated disc and significant pain. The phrase *slipped disc* is a bit of a misnomer; nothing actually slips in the process. The discs more ooze than slip, but I suppose oozed disc does not have quite the same ring to it.

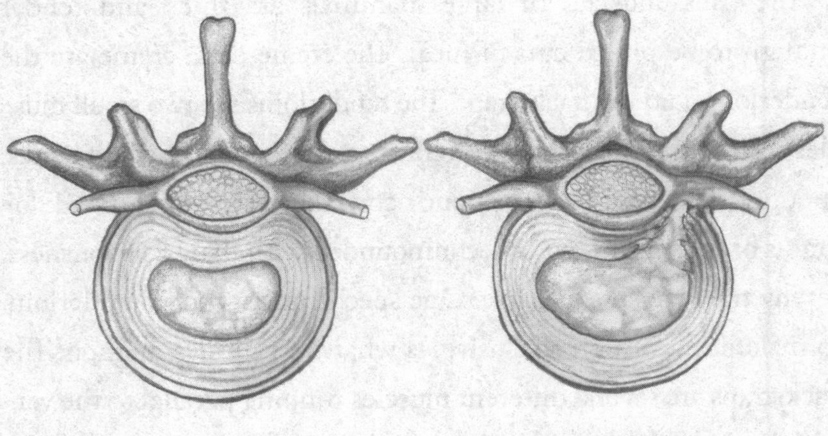

Happy disc **Sad disc**

THAT'S NOT A KNIFE

It is hard to appreciate just how many muscles there are in the back. There are layers upon layers of muscles attached to the vertebral column to allow it to flex, extend, and rotate. In addition to being numerous, the back muscles are also quite meaty. I remember walking into the cadaver lab one time and seeing one of my best students wielding a large hunting knife. He had spent hours with a small scalpel trying to cut through the back muscles to expose the spine. He eventually took measures into his own hands and brought some heavier equipment into the lab. We had already finished the muscle identification portion of the class, so he only needed to reflect the muscles out of the way so he could access the vertebrae. It worked like a charm, and as a result I added an eight-inch hunting knife to the cadaver-lab tool drawer. In Idaho, most of the students are already quite comfortable with a Crocodile Dundee type of knife.

The back muscles of large mammals are thick and tender enough to be prized cuts of meat. The crème de la crème are the tenderloins and the backstraps. The tenderloins are two small muscles connecting the lumbar vertebrae to the femurs, or thigh bones. In a 500-kilogram (1,102 pound) cow, there are only a few kilograms of tenderloins, which, compounded with their deliciousness, is why they are so expensive. One specific section of a tenderloin, to make it even more exclusive, is what we call filet mignon. The backstraps are several different muscles running parallel to the vertebral column along most of its length. In quadrupeds, tenderloins and backstraps do not have to work terribly hard. When muscles work and contract all day long they get dense and tough. Most of the muscles attached to the weight-bearing portion of the skeleton are ground up into hamburger. When muscles sit around without much to do, they become very tender. Those cuts are pricier, and we give them names like prime rib and T-bone.

Up on two feet, the spine has to support more of the weight of the body (which is basically the whole problem), and thus, back muscles have more work to do. Strong back muscles make for less of a strain on the discs and the vertebrae. Importantly, there are other core muscles that also play an integral role in the health of the spine. Keeping those abs strong can greatly relieve the stress on the lower back, and it makes you look sexy as an added bonus. I'm going to go do a set of crunches before carrying on.

THE COUCH POTATO HYPOTHESIS

Shifting the focus from nature to nurture, a counter-idea to the ancestral shape hypothesis focuses on muscle strength and flexibility. The sedentary lifestyle hypothesis posits that back pain occurs in epidemic proportions because we do not put our backs, with all their interconnected bones, muscles, ligaments, and tendons, through the same historical paces we once did. It suggests spines become misaligned and discs slip because our backs are not flexible and strong enough owing to general inactivity.

There are piles and piles of research articles from studies

exploring the issue of lower back pain. It makes my back hurt just thinking about all the people who have been subjects in back-pain research. I swear my back has started to hurt more while working on this chapter.

Some of the back-pain research focuses on anatomy (like the aforementioned research on vertebral shape), but more of the studies zero in on behavior and how it relates to back pain. With hundreds of studies, as you might expect, the results are highly variable. Some of the evidence gathered supports the idea of physical activity preventing lower back pain.[14] Other research presents the opposite result, suggesting physical activity is a risk factor for lower back pain, especially if the activity is overly strenuous, with consistent elements of lifting, twisting, and bending.[15]

Still other surveys have bridged the gap with data suggesting a U-shaped relationship between physical activity and lower back pain.[16] Too little activity may lead to pain and too much activity may also lead to pain. Play a reasonable amount of golf and your back can stay flexible, healthy, and limber. Play too much golf and you're Tiger Woods, who has had multiple back surgeries and has had times in his life when he needed help getting out of bed.

One consistent element found in articles about back pain is the nature of people's employment. Not surprisingly, no matter how good your lifting technique, if you have to lift heavy objects all day for your job, it is eventually going to wear you and your lumbar vertebrae down.[17]

For the average deskbound Joe, despite a lot of contrary headlines in the popular press, there is scant evidence that sitting at work causes lower back pain.[18] Sitting all day definitely is not good for you (it has been linked to problems like high blood pressure and heart disease), but doing anything for the whole day without taking breaks is not good for you. It is trendy to call sitting the new smoking and standing desks have exploded in popularity, but the latest evidence suggests standing all day is just as bad as sitting all day, if not worse.[19] The trick is to mix it up.

SAVE THE EARTH—BRING BACK THE PIRATES!

It is difficult to get clear answers about the causes of back pain because so many of the back-pain studies suffer from the same problematic experimental design. Most of the research focuses on subjects who already have back pain. Just because someone is not physically active and has a sore back does not mean a sedentary lifestyle caused their sore back. It may, in fact, be the opposite. People with a sore back may be inactive because they have a sore back. It is nearly impossible to tease those two variables apart if a study is purely correlational. The notion that correlation does not imply causation is one of the most important principles students can learn in an introductory statistics class.

Multiple sets of unusual data stress the point about correlation and causation. A classic example I use in my classes demonstrates a very strong negative correlation between the global temperature and the number of active pirates on Earth. As the number of people employed in piracy around the globe has dipped in the last

200 years, there has been a steady climb in the earth's temperature. In spite of the correlation, you would have to be a blithering idiot to suggest that one is the cause of the other.

Another great example is the correlation between ice cream sales and rates of violent crime. More violent crime occurs during the summer months, and those are the same months when ice cream sales increase. If correlation automatically meant causation, then we could blame the uptick in summer murder rates on ice cream and its deceptively delicious way of throwing people into a murderous brain-freeze-induced rage. You know the day is coming when some politician sincerely suggests we could control crime by banning ice cream and stop global warming by bringing back more pirates. Arrrgh!

Many back-pain researchers struggle with the correlation/causation conundrum because they rely on survey data in which individuals with back pain are the research population. It is logistically more difficult, but back-pain researchers need to divide participants into behavioral groups early in their lives *before* they ever develop back pain. If we really want to know what causes back pain, we need to start off with people who do not have back pain. Once they have back pain it completely changes how they live their lives, and the data become very muddy, very quickly. All we need to do is convince a bunch of 12-year-olds to be part of a 50-year study exploring the roots of lower back pain. The global issue of back pain literally costs society trillions of dollars per year, so we should be willing to throw some money at the problem. Offer kids a new phone once a year as long as they remain part of the study. There would be, like, a zillion tweens lined up immediately.

ONE DEGREE OF SEPARATION

Back pain is sneakier than other issues that cause pain or discomfort. It can ebb and flow more than problems like foot pain, crooked teeth, or blurry vision. If your eyes are lousy, then they're lousy, and there is no getting around that. If standing on your feet all day makes your feet hurt, that is not likely to change. Crooked teeth do not just straighten themselves out for a few weeks and then revert to being crooked. Back pain can be different. It might creep up one month and then magically disappear the next. It can go away for years and then come back with a whole different degree of severity than previous iterations.

Even if not afflicted themselves, most people do not have to look far to find someone they know who suffers from severe back pain. It is more of a one-degree-of-separation condition rather than a six-degrees-of-separation condition. My back pain is not bad, yet, but I work with three people who have had back surgery, and my primary care physician, Dr. Martin, has dealt with brutal back pain in recent years.

Dr. Martin first dealt with back pain during his residency. I am betting the total lack of sleep and hours upon hours on his feet did him no favors. An MRI showed early signs of disc degeneration in his lumbar spine. He added some stretching to his daily routine, and the pain mostly went away. It crept back occasionally but never in a way that significantly changed his quality of life.

The pain returned years later when he became overloaded with work and was unable to exercise and be active in the same ways he always had. He believes that remaining active, with a strong core, is the ultimate key to avoiding back issues. The proper shape of the spine is lost once the muscles attached to the spine weaken from inactivity. The discs are then placed under even more intense pressure than their typical situation, which is already rather labored owing to the whole walking-around-on-two-feet issue.

Sometimes the lack of activity does not directly cause pain but can put the back in a more injury-susceptible state. This is what happened to Dr. Martin. We had a terrifically hard winter a couple of years ago, with layers of snow and ice on the ground for months on end. In one of the most classic ways to get hurt, he slipped and fell on a patch of ice and injured his back. The injury was not debilitating, but it was bad enough that it inspired him to try to develop a stronger core. He ended up layering a second acute event on top of the first when he felt a pop in his back during a session of yoga.

The yoga mishap was in January. Over the next few months, the pain continued to build in intensity. Imaging revealed that his problem was in the same location it had started in his 20s, with a bulging disc between his third and fourth lumbar vertebrae. His annulus fibrosus (the ring holding in the gel in the middle) was completely torn on one side, and the nucleus pulposus had squeezed out and was pushing on nerves. He got an epidural in an attempt to relieve the pain without surgery. Being a physician, Dr. Martin knew back surgery is a last resort, an option that should be performed only if a patient is crippled by pain. It is still unclear how effective surgery even is at eliminating back pain. For some patients, the risks of complications like infections and nerve damage and the possibility of induced dependence on post-op pain medications may outweigh the potential benefits. By August, however, the pain was unbearable and surgery became a choice Dr. Martin had to consider.

Most back surgeries follow one of two paths. In a microdiscectomy, surgeons carve out the bulging section of disc, creating space for the remaining tissue. By most reports, a microdiscectomy leads to a good or even excellent outcome upwards of 90% of the time. The alternative is to have spinal fusion surgery. In spinal fusion surgery, a surgeon removes the entire problematic disc and fuses together the vertebrae from above and below the removed disc, typically using pieces of bone harvested from a hip of the patient. As you might expect, the second route is a far more serious procedure.

Dr. Martin started with a microdiscectomy. It did not work; he was part of the unlucky 10%. He had another microdiscectomy a month later to remove more of the bulging disc, and again, it afforded him no relief. Worse still, the pain went to a whole new place. It became "uncontrollable, mind-boggling pain," and he needed to use a wheelchair for portions of the day. The next month he underwent spinal fusion surgery, his third back surgery within a span of three months. The spinal fusion surgery was, at least to date, successful. The clinic sent his excised disc to the pathology lab, and the results showed the problematic disc was infected. He had contracted an infection during the second surgery, and the severe infection layered on top of the ruptured disc is what caused the pain to ramp up to an unendurable level. A month of painkillers and two months of antibiotics later and things were looking better. Now, more than a year after surgery, it looks like Dr. Martin may have come out into the light at the end of a long back-pain tunnel.

One positive outcome is that Dr. Martin can make a very convincing case to his patients for maintaining the strength and flexibility of their cores before it is too late. He has made it his new life goal to keep as many people out of back surgery as possible. Patients now leave his office acutely aware that their backs need to be strong *before* the slip on the ice or the tweak during an exercise class. It is one thing to hear that kind of advice from someone who has never experienced back pain or from someone who has had back pain but has no medical training. It's a whole other thing to hear it from a trusted, reliable physician who had his life turned upside down by unrelenting, unimaginable, and inescapable pain.

A GOOD NIGHT'S REST

When I sat down with Dr. Martin to talk about his experience, one of the last things we discussed was the effect of back pain on sleep. He has had to completely reinvent how he sleeps in response to his problems of the past few years. To bring the back-pain story full circle, let's put in the earplugs, strap on the eye mask, and get down to the business of how humans sleep. With somewhere around one-third of our lives spent zonked out and so many people struggling to sleep because of back pain, it only makes sense to take a critical look at how humans catch their ZZZs.

Exploring how other great apes sleep sheds some light on how humans should approach getting meaningful rest and how to avoid back pain while sleeping. We are a unique primate in terms of how we sleep. Namely, we sleep on the ground, which is downright unusual for a primate. Great apes tend to sleep, like other primates, up in trees. The non-great-ape primates tuck into a nook or find a comfortable branch to curl up on. Sleeping on a branch works fine for a little lemur or a tarsier, but it does not work well for an animal as large as an orangutan or a chimpanzee. They might crash to the ground if they try to stretch out on a tree limb.

The solution for large-bodied tree-sleeping apes is to make a bed. Each day, chimpanzees and orangutans gather materials to make their beds for the night. It's like staying at a hotel where you always sleep in a freshly made bed. The only differences are that the bed is in a tree and it is made out of branches and leaves. Also, the maid is an orangutan who makes the bed and then sleeps in it. Other than those things it's the same. It does make me feel better about being a sleep diva knowing it is somewhat ingrained in the DNA of hominids.

There are advantages to sleeping up in the trees. Namely, there are fewer predators and fewer biting insects, and it is warmer than the ground. For the great apes, beds carry the additional benefit of being comfy. The notion of comfort has led researchers to develop the sleep quality hypothesis in which they propose that the added comfort of beds allows great apes to reach a deeper, heavier state of sleep than is possible when trying to balance on a branch.[20]

Researchers snoozing out in the forest with the chimpanzees

made an interesting observation about chimpanzee beds. They sampled 1,844 chimpanzee night beds (take a second to appreciate that large sample size) and discovered that chimps used the same type of tree to build a nest in 73.6% of cases. Interestingly, the preferred tree made up only 9.6% of the trees in the forest. In other words, the sleepy chimps were not grabbing branches at random and knocking out shoddy, makeshift beds. They were being selective about their mattress materials.

The sleep researchers also analyzed the properties of the preferred trees. In the article they published in the journal *PLOS ONE*, they note that the most coveted type of tree was a species of ironwood that "was the stiffest and had the greatest bending strength" of all the options for bedding materials available to the chimps. So chimps go for mattresses with some give, but ones that are also stable and firm.[21]

Even after the dawn of bipedalism, humans likely still scrambled back up into trees to sleep. There were still all those pesky predator and insect problems on the cold, hard ground. Of course, humans eventually moved the nighttime operations down to the dirt entirely. A critical step in the trend toward terrestrial sleeping was the control of fire. Fire kept the bugs at bay and solved the temperature problem. Humans are social animals, and sleeping in a group allowed for a high level of vigilance against large predators. In a big enough group, there is usually someone awake at 3:00 a.m. to warn everyone when a toothy predator comes into camp.

Ground sleeping around a fire took sleeping to a whole new level. Early humans began to make beds out of cushy materials like grasses and sedges.[22] Sleeping on mattresses around a fire in a social group likely led to an even deeper, heavier sleep. Some have gone so far as to propose the sleep intensity hypothesis, in which they advance the idea that intense, high-quality sleep allowed our human ancestors to take a great evolutionary leap forward.[23] Human sleep became the deepest, most peaceful rest a primate had ever experienced. Better sleep translates to needing less sleep, and the additional active time can be put to good use learning new skills. New skills separate humans from all other animals, and as long as cotton-top tamarins and mouse lemurs are snoozing so much, there is zero chance they are going to take over the world. It's too bad, because it would be totally adorable if mouse lemurs were the face of civilization.

GETTING BACK TO THE BACK

Tree branches, grasses, and sedges aside, the question remains about which type of modern mattress is best for back pain. There is a prevailing opinion that a stiff mattress is better. When orthopedic surgeons were surveyed about the issue, the vast majority thought mattress choice was a critical part of lower back pain management, and 76% believed a firm mattress was the right choice for those with lower back pain.[24] Those opinions, however, were not sufficiently grounded in research, so a group of Spanish scientists set out to tackle the question head on with a randomized double-blind trial.

Before and after their siestas, the researchers gathered up hundreds of adults with chronic lower back pain and randomly assigned them to either firm or medium-firm mattresses, based on a firmness scale developed by the European Committee for Standardization. I had no idea mattresses had scores, but at least in Europe, the scale is from 1 to 10, with one being as hard as a rock and 10 being extra fluffy. The firm mattresses used in the study were a 2.3 on the European scale, and the medium firm were a 5.6. The study did not have a soft-mattress category. If my wife, Julie, had been in charge, she would have included something up there in the 8 to 9 range.

The randomly divided subjects did not know which mattress group they were a part of (the first blind). The doctors performing the patient assessment after the 90-day trial did not know which mattress group the patients had been part of (the second blind). When possible, clinical trials should always be double blind, because double-blind studies eliminate the possibility of bias on the part of both the patients and the doctors.

Maybe the Spaniards did not include soft mattresses because they were so sure the firm mattresses were going to be the preferred choice. Turns out, not so much. In an article published in *The Lancet*, they reported that "at 90 days, patients with medium-firm mattresses had better outcomes for pain in bed, pain on rising, and disability than did patients with firm mattresses."[25] The chimpanzees had it right all along. You want something firm, but also with a little bit of give. I am not sure where that leaves Julie and me for the purchase of the next mattress. Maybe we should go super old school and try building a nest up in a tree.

TAKE A HOT SHOWER

Of all the issues covered thus far in the book, back pain seems to have the most potential to ruin quality of life. There are relatively easy fixes for lousy teeth or bad eyes. Nobody likes to have to get braces or need glasses, but those are simple and effective solutions compared with spinal fusion surgery or the long-term use of opioid medications. As long as you don't die from choking (which admittedly has a very negative impact on quality of life), the intersecting trachea and esophagus is a rather manageable evolutionary quirk. There is no doubt having sore knees or sore feet can be miserable. But when your day is over, you can get off your feet. Knee and foot pain are unlikely to persist around the clock. In contrast, there is no hiding from back pain. It can affect quality of life 24 hours a day, 365 days a year.

More than any other issue covered to this point, the back is going to get you. It is, after all, the leading cause of disability worldwide. If you are unlucky, the pain starts in your teens or 20s, maybe in relation to an athletic endeavor or some act of youthful hubris. More typically, the pain will hold off until your 30s or 40s. If you are incredibly fortunate, it may not even kick in until after 50. But it is coming. There is no avoiding it. The weight of your body squashes those discs every day, and there's not a damn thing you can do about it. You can lose weight, strengthen your core, and not help your friends move their sleeper sofa, but eventually the checkered evolutionary past of humans is going to catch up to you. At the end of the day, you're a bipedal beast walking around with largely quadrupedal parts.

Sorry. Schedule a massage, take a hot shower, and try to remember the great benefit that hopefully offsets the cost of the pain—your hands are free to do whatever you want. All those other critters may not suffer from as much back pain as humans, but they are also stuck scrabbling around on all fours. I do think we got the better end of the bargain, but I also don't suffer from excessive back pain. If I ever have a herniated disc, you can bet I'll be cursing the bipedal life all the way to the doctor.

PART THREE

Bundles of Joy

7

To Bleed or Not to Bleed

There are five liters of blood in the average adult. How much blood does the average woman lose via menstruation over the course of her lifetime?

a. 3.2 liters
b. 12.9 liters
c. 23.7 liters
d. 72.4 liters

So I was buying tampons the other day. This does not occur on a regular basis. Julie keeps her waterfowl very collinear, so there are typically more than adequate quantities of feminine hygiene products in the house. Once every blue moon, however, the circalunar joy of menstruation sneaks up at a time when the supply has dwindled. When this happens, as any good husband would, I drive off into the dark of night to try my luck at buying tampons and pads.

Julie is smart about this. She does not send me in blind. Lord knows what I would come home with. I would probably forget why I was at the store in the first place and come back with some chicken wings and a six pack of pop. I go in with a specific goal, including, these days, a picture of the desired products on my phone. I also go with a plan B and a plan C, since the preferred options are often out of stock.

It sounds simple enough on paper. It shouldn't be any harder, really, than buying a box of cereal. Get the Wheaties, and if they're out of Wheaties, buy the Honey Nut Cheerios. But, sweet buttercup, the tampon aisle is nothing like the cereal aisle. I always forget just how many options there are going to be. Somehow, in my mind, I will walk in and there will be only one type of tampon and one type of pad. They will be in stock, I will choose a couple of boxes that don't look like they were drop-kicked, and I'll be back home in a flash.

If you are a guy who has not had this experience, let me assure you it works nothing like that. There are dozens, maybe hundreds of options. And unlike the boxes of cereal, they all basically look the same. Every box is dark and has small, neon letters. Their outer similarities belie their degree of variability; the contents of each package are wildly different.

For example, choosing a product at random, one of the available pads for purchase is Always Ultra-Thin Fresh Size 4 Overnight Pads with Wings, Scented, 24 count. Not counting the word *Fresh* (it seems unlikely there would be an unfresh option), there are seven variables. The brand (Always), the thickness (Ultra-Thin), the size (4), the use (Overnight), the shape (with Wings), the smell (Scented), and the quantity (24). If your poorly stocked box store is anything like my poorly stocked box store, you're doing well if you can get four or five of those seven variables. Six would be a spectacularly good day, and hitting all seven is like winning the lottery. And then, of course, rinse and repeat with just as many options for the tampons. Good luck. I guess this is why Amazon is taking over the world. You can get all seven variables, every time. What you cannot do is get your order in 30 minutes . . . yet.

MENSTRUATION TERTILES

The tampon-purchasing experience always leaves me with the impression that menstruation must be the most variable of all human traits. A 2012 study with the least flowery title of all time, "Menstrual Bleeding Patterns among Regularly Menstruating Women,"[1] charted the menstrual behavior of 201 women for two consecutive cycles. As you might expect, based on the tampon aisle, the results were highly variable. All of those different menstrual product options exist for a reason: supply matches demand. The median cycle length for women in the study was five days, but the range was 1 to 16 days! Some women zip in and out of their periods in less than a weekend, and for others, the experience lasts more than two weeks. If I were a woman whose period lasted as long as the Olympics, I would find the speed menstruaters super maddening.

The researchers organized the menstrual flows into light, medium, and heavy tertiles (a new word for me—it's the same idea as quartiles or quintiles but with three groups instead of four or five). As with the length of a period, the amount of blood lost during one cycle was highly variable. Light cycles were ones in which women bled fewer than 36.5 milliliters (ml). The medium tertile was for cycles between 36.5 and 72.5 ml of blood loss, and the heavy tertile was for cycles where there was more than 72.5 ml of blood lost. For the one-third of cycles defined as light, the average amount of blood lost was 15.2 ml.

To make those numbers more meaningful for those who grew up in one of the three remaining countries where the government refuses to use the incredibly logical metric system (I'm looking at you, Burma, Liberia, and the US), 15 ml is roughly three teaspoons of blood. For the one-third of cycles defined as heavy, the average amount lost was 114.4 ml. That is almost half a cup of blood. And that was the *average* in the heavy cycle category (I can only say tertile so many times). So there were plenty of women with cycles involving blood loss well north of half a cup. Looking at the averages, a heavy cycle involves 7.5 times as much blood loss as a light cycle. And you know the women with the 36-hour periods are also the same lucky ones who bleed only three teaspoons each month.

The average amount of blood lost, across all types of cycles, was 45.5 ml, or just shy of a quarter cup. A quarter cup of blood is not a small amount. If you were chopping carrots for dinner, sliced your thumb instead of the veggies, and a quarter cup of blood came out, you would rush to the ER. If you work the math using an average cycle length of 28 days (in another study with more than 2,000 women, the average length was 28.1 days[2]), over the course of a year it comes out to 593 ml of blood. Over roughly 40 years of menstruation, that adds up to 23.7 liters (more than six *gallons*) of blood lost over the course of the reproductive years. That is a staggering amount of blood loss. There are only five liters of blood in the average adult. A woman loses the equivalent of all the blood in her body almost five times over. And remember, these are averages. There are plenty of women out there losing twice or even three times the typical amount of blood.

The researchers did the whole tertile thing so they could dig into menstruation demographics. The story there is, try to enjoy the good times while you're young. Older women bled more than younger women in the study. The women in the study tended to be young (participants had an average age of 27.7 years), which means, if anything, the estimates of blood loss are likely conservative. Also, interestingly, women in the medium and heavy menstruation tertiles had a statistically significant earlier age of menarche (first menstruation at 12.5 years, on average, for both categories) than women in the light menstruation tertile (menarche at 12.9 years, on average).

Such a significant loss of blood can lead to a serious iron deficiency in women. Remember the study on global causes of disability I referenced in the last chapter? It was published in the British journal *The Lancet* and funded by the Bill and Melinda Gates Foundation. It identified lower back pain and major depressive disorders as the two leading causes of global disability. Do you recall the next leading cause? It was iron-deficiency anemia.[3] In fact, there are several countries in which iron-deficiency anemia is the leading scourge of the people. The issue is of particular concern in sub-Saharan Africa, where, in many cases, the available dietary options do not provide access to good sources of iron. The problem did not even crack the top 10 causes of disability in developed countries, falling behind other issues like anxiety, falling, and COPD. But in developing countries (and overall globally), it slotted in directly after back pain and depression, meaning it causes more people continued trouble than major issues like migraines, diabetes, and neck pain.

Iron-deficiency anemia is no walk in the park. Iron is a vital component of hemoglobin, the molecule oxygen binds to in red blood cells. Without adequate iron, the body cannot make adequate hemoglobin. Without sufficient levels of hemoglobin, it is not possible to deliver the necessary amount of oxygen to the tissues of the body. Critical physiological processes start to slow, inducing symptoms that include fatigue, weakness, headaches, and shortness of breath.

There is another, slightly more out-there symptom associated with iron-deficiency anemia: unusual cravings. The body knows when it is running low on a critical ingredient and it searches out the missing nutrient. There are historical records of this happening with something as plain and simple as salt. If the body is unable to absorb and process salt in the typical manner, an intense craving for salt can occur. A story about this happening to a kid in the 1930s was published in the journal *Endocrinology* in 1940.[4] The little boy lived until he was three years and seven months old, and during his short life he consumed extra salt in any way possible. He would accept only salty foods (he was a big fan of salt mackerel—it was a different time), and he would sneak plain salt from his parents whenever he could get his hands on it. He had a raft of other endocrine-related (hormonal) medical issues that eventually led to his hospitalization. Unfortunately, the hospital placed him on a standard diet. Unable to forage for the copious amount of salt his body craved (and needed), he died within a week.

Other nutritional deficiencies may cause highly unusual cravings. In some cases, they may lead to cravings that in no way help the body build up the supply of the missing ingredient. It's like the body is searching for the missing nutrient but has no idea where to find it. Serious deficiency of an essential element, like iron, may cause the afflicted to crave and eat wholly non-nutritious food items. We are not talking non-nutritious like candy bars and potato chips. These are *seriously* non-nutritious objects like cigarette butts, rubber bands, and chalk. The phenomenon of ingesting non-nutritious items is known as pica. Pica remains a bit of a mystery for the medical community. One of the most commonly craved and ingested substances in these situations is dirt. There are many ideas on the table for what causes pica (including mental illness and stress), but an article published in the *Journal of Medical Case Reports* suggests "it seems to be strongly associated with iron deficiency anemia, and in the majority of cases the unusual eating and behavior disappears upon iron supplementation."[5]

Women of reproductive age are particularly susceptible to iron-deficiency anemia because of menstruation. In a study of a Danish population, premenopausal women had a 10-fold higher rate of iron-deficiency anemia than did men.[6] And, mind you, this occurred in a country where most people have easy access to the meat, leafy green veggies, and iron-fortified cereals necessary to replenish iron stores. The Gates-funded study did not break up the data by gender, but I would bet the issue is particularly glaring among the women in sub-Saharan Africa, where access to adequate dietary iron is a glaring health issue.

The problem of iron-deficiency anemia effectively goes away after menopause, further evidence that menstruation is the primary culprit. Zero of the postmenopausal women in the Danish study group had the condition.

COVERT OPERATIONS

Before we get too deep into this, we need to take a minute and define what it means to menstruate. It's a little more complicated than it might seem. Placental mammals build up and break down the inner linings of their uteruses in a cyclical manner. What happens to the products of the breakdown is where there are two distinctly different paths. Some animals reabsorb the built-up tissues and have no visible, external signs of bleeding. Scientists refer to it as covert menstruation. Other animals do not just shed the inner lining of the uterus, known as the endometrium, but also discharge it from the vagina. Technically this is overt menstruation, but for all intents and purposes, it is the behavior being described when the word *menstruation* is used without any qualifiers.

The discussion of the roots of menstruation takes a unique turn when comparing different mammals. I am always thinking about what I want my students to remember in 10 years. I joke with them about a hypothetical 10-year class reunion and how there are some topics I expect them to still remember and understand a decade after finishing my class. They may have forgotten some of the details, but I want them to hold on to those big ideas forever. Here is a core idea for this chapter. Please remember it for the 10-year reunion of everyone who reads this book. All three dozen of you can get together in a decade and quiz each other. I'm going to put it in bold and all caps with excessive punctuation, so it really smacks you in the face. Ready? Here it comes: **MOST MAMMALS DO NOT MENSTRUATE!!!!!!** Most mammals reabsorb the excess blood and tissues, like a sponge soaking up extra fluid.

Take a minute to let that sink in. It is quite the revelation for most people. The most authoritative published list of mammals includes a total of 5,416 species.[7] The only ones who menstruate are the 19 apes (including us), 132 old-world monkeys, half a dozen or so new-world monkeys (there are still gaps in the menstruation data for new-world monkeys), four species of bats, a rodent called the spiny mouse, and elephant shrews.

It's hard to believe someone got all up in the business of elephant shrews and determined that they menstruate, but that's science for you. If you include all 19 species of elephant shrews, it adds up to 181 menstruating species out of 5,416 total mammals. Even if you throw in an extra 100 menstruating mammals in which menstruation has yet to be discovered (and I am betting an extra 100 is a liberal estimate if we already have it figured out in random species like leaf-nosed bats and elephant shrews), the behavior is still limited to only 5% of mammals.

For all you dog lovers, do not confuse a little blood coming out of Fifi with menstruation. Any blood coming from down there in a female dog is a result of a vaginal discharge associated with coming into heat. It is not from menstruation. Just like most mammals, dogs reabsorb their menstrual blood. In some of the mammals with estrous cycles (like dogs), the reproductive tissues become engorged with so much blood when they are sexually receptive that excess blood ends up getting discharged. Others, like cats, keep the process tidier (cats always keep things tidier) and pass a clearer, more watery discharge.

CURSE OF THE MODERN LIFE

Before we get to the big "why do women menstruate" question, let's start by making it clear that, when it comes to menstruation, modern women have it far worse off than their predecessors. For starters, they have many more menstrual cycles than human females did historically. Once hitting reproductive age, our ancestors would have started having kids. None of this go to high school, go to college, meet the right partner, start a career, buy a house, get a dog, have one kid, maybe two, and call it good. Starting in their teens, our ancestors would have begun a pattern of get pregnant, give birth, nurse for two to three years, get pregnant, give birth, nurse for two to three years, get pregnant, give birth, nurse for two to three years. Rinse and repeat 8, 10, or maybe even 15 times. There was very little down time between being pregnant and nursing in which to have a menstrual cycle. To get a sense of just how many menstrual cycles our ancestors might have had, it is most helpful to find current groups of people still reproducing in the traditional, natural fertility manner.

Beverly Strassmann is a professor of anthropology at the University of Michigan and has studied such a population in Mali, West Africa, for more than 30 years. She has looked at many aspects of the reproductive behavior of the Dogon of Mali and has shed light on what menstruation and child-rearing may have looked like historically. The Dogon have strict cultural norms by which menstruating women are segregated from the rest of the village during their cycles.[8] This behavior (which Strassmann has discovered is a male-driven practice that easily allows men to ascertain which women are close to the fertile portion of their cycles) has allowed for straightforward data collection about menstrual cycles. Dogon women experience, on average, roughly 100 total menstrual cycles in their lifetimes (the mean in the study was

109 and the median was 94) and birth, on average, 8.6 children. Those numbers are strikingly different from what women experience who are not practicing natural fertility. Strassmann estimates, based on data from other researchers, that it is not unusual for modern American women to go through as many as 400 menstrual cycles in their lifetimes.[9]

To further complicate the issue for modern women, menarche occurs much earlier now than it did historically. Before the industrial revolution, a young woman could expect to have her first period sometime deep into her teenage years. The average age of menarche in 1840 was 16.5 years.[10] These days it occurs, at least in industrialized societies, at more like 12 to 13 years of age, and it is not unheard of for a girl to have her first period before her 10th birthday. Many people have speculated about why menarche is occurring earlier in modern times, and there are two leading hypotheses. The first points the finger at environmental contaminants. Industrialization has caused a dramatic increase in the quantities of environmental pollutants, which, ultimately, make their way into our bodies. Some of these pollutants interact with and disrupt the normal operations of the endocrine system. Some have suggested that early and repeated exposure to contaminants like bisphenol A (BPA) has contributed to the dramatic decrease in menarchal age.[11]

The other idea has to do with fat deposition. Industrialization has put more food on the table and, at least in many locations around the globe, women's bodies are able to get into a reproductively ready state at a much earlier age than ever before. The decrease in the age of menarche may simply reflect an increase in access to adequate and varied nutrition. The age of menarche has continued to fall as the global diet has changed. Eventually, the diet went well past the point of simply being adequate and diverse to there being a burger joint on every street corner. In recent years, the decreasing age of menarche has finally started to level off, which may indicate the world has become saturated with fast food restaurants, or it could be that we have simply hit the biological limit for how early girls can enter puberty.

While it will be difficult for researchers to tease these two ideas apart (understanding they are in no way mutually exclusive), data from the study of the Dogon women strongly suggest the culprit is, in some way, modern, industrialized culture. The median age of menarche for the women in Strassmann's study was 16 years, not notably different from the numbers reported for women in the 1840s.

Not to be forgotten in a discussion about the decline in the average age at menarche is the recognition that women can become pregnant at younger ages. Having menstruation start at a very young age is admittedly not ideal, but taken by itself it likely does not cause a dramatic change in the arc of someone's life. Becoming pregnant at 13 is an entirely different story. It was not even biologically possible for most women prior to the Industrial Revolution. It is very much a possibility now. I met a 30-year-old a few years ago who had just become a *grandmother*. She had her daughter when she was 17, and then her daughter went through puberty at 13 and got pregnant on her first ovulation. It is common for the first few cycles of a woman's (or girl's) life to be anovulatory (no egg produced), but it is not a guarantee. A period happens only when reproductive hormone levels decline after an egg is ovulated and then not fertilized. If the first egg is fertilized, parents don't even get the cue of menstruation to know it is time to have the talk.

THE BIG QUESTION

We cannot put off the question any longer. Why do human females menstruate? How did a system leading to significant blood loss, significant iron loss, and a significantly lousy few days every month ever evolve?

For every issue covered up until this point, before writing the chapter, I had a very solid idea of the hypotheses and the evidence I was going to present about the origins of each ache and pain of the human body. I understood the evolutionary history of our jaws and brains and how they help explain why our teeth do not fit in our mouths. I knew enough about the evolution of the eyes in vertebrates to explain why they do not always work perfectly. I could explain why the trachea and esophagus must intersect, leading to the problems described in the chapter about the throat. It is not exactly a huge mystery that the human skeletal system has some issues because we evolved from quadrupedal animals and now make our way around on two feet. Those bipedal complications were the foundation of the chapters on the knee, foot, and back.

The work was in organizing the experiments and evidence and figuring out the most compelling way to present all the fascinating research on the subjects.

With the topic of menstruation, it has been an entirely different experience. Why do women menstruate? Before I did the work for this chapter, I had no idea. Don't get me wrong: I understood, physiologically, why menstruation happens, but I did not have the foggiest clue as to how or why it evolved. I am not one bit ashamed of this because, it turns out, it is wildly complicated and still not terribly well understood. Even the people who study uteruses for a living are only just beginning to sort out the answers.

RISE OF THE PSEUDOSCIENTISTS

Coming up with hypotheses has never been the problem for the menstruation research community. There just has not been a lot of evidence to support the hypotheses. For decades, researchers worked under the backdrop of significant cultural taboos about menstruation. Also, until deep into the 20th century, men did nearly all the research, which I'm sure did not help in terms of moving past some of the preconceived stereotypes.

The stigmatization of menstruation has been around for a very long time. There is a weirdly lengthy and detailed description of menstruation rules and regulations laid out in the Bible in the Book of Leviticus. It is chock-full of ridiculousness, but the most ludicrous part of the passage is the requirement that at the end of her cycle, a woman is supposed to "bring two turtledoves or two young pigeons and present them to the priest at the entrance of the tabernacle. The priest will offer one for a sin offering and the other for a burnt offering. Through this process, the priest will purify her before the Lord for the ceremonial impurity caused by her bleeding."[12] What a nightmare that must have been. After each cycle, women were supposed to run around and try to capture some birds? Have you ever tried to catch a bird? It is not exactly trivial. If anything is making a woman dirty, it is not her cycle but rather trying to catch and transport two live birds to church to have them killed. The whole thing makes going to the store to pick out tampons seem like a walk in the park.

Building upon that very flawed foundation, the idea of menstruation as a symbol of impure and unclean women picked up new steam in the early 20th century because of anecdotal evidence involving flowers. A Hungarian-born American physician, Dr. Bela Schick, began his dive down a deep pseudoscience rabbit hole when one of his nurses told him she caused flowers to wilt if she held them during her period. He ran with this observation and invented the concept of a menotoxin—a powerful, toxic element of menstrual blood that was inherently poisonous and could not only cause flowers to wilt but also keep bread from rising and jam from setting. Getting jam to set was very important in the 1920s, so these vicious jam rumors had to be taken seriously.

Thus, we hit upon one of the first ideas for why menstruation occurs. The research community believed that, in order to purify the female body, the vile female toxins had to be regularly flushed from the system. Dr. Schick and others expanded the idea to suggest the presence of toxins in all the fluids of a woman's body, from menstrual blood to sweat and breast milk. Male fluids were all conveniently ignored.

Scientists ran a number of experiments in the ensuing decades, all performed with researchers strongly wed to a preconceived idea of how they wanted the results to turn out. They were single-mindedly intent on proving the existence of menotoxins. Doing science through a biased viewpoint leads to shoddy outcomes. When researchers are invested in the results coming out a certain way, they skip controls, don't conduct experiments in a blind manner, and don't publish neutral or negative results. Any data that do not support their preconceived notions get shoved deep down in filing cabinets (or these days on external hard drives) and are never published.

Once scientists explored the idea of menotoxins through a rigorous, critical scientific lens, the concept crumpled. Women do not have poisonous elements in their bodies, at least not to any greater degree than men do. Women do not menstruate to purge some inherent toxic element found exclusively in women; menstruation simply eliminates unused blood and tissue. The menotoxin concept had built up some inertia, however, and as a result, it took the better part of the 20th century for the notion to be wholly discredited.

With the idea of menotoxins finally dead and buried, the question was back on the table for why menstruation exists. Why is there so much unused blood and tissue in the first place? It was time to test some new ideas.

THE SPERM GET A TURN

Before we get to the most modern hypotheses, there are a couple of other ideas deserving of attention. In 1993, evolutionary biologist Margie Profet put forth a different notion for why women menstruate, placing the blame on the other side of the coin: dirty sperm. She titled the paper "Menstruation as a Defense against Pathogens Transported by Sperm," and it made quite a splash in the popular press.[13] The anti-pathogen hypothesis she introduced suggested bacteria from both the male and female sexual organs hijack their way onto sperm and get a free ride up into the warm and moist uterus. Bacteria love nothing more than warm and moist, so this is obviously a win for them. Profet suggested menstruation evolved as a way of cyclically flushing out invasive bacteria in order to keep the reproductive environment free of pathogens.

The anti-pathogen hypothesis generates several testable predictions. It predicts there should be fewer uterine pathogens after menstruation than before and, also, that there should be a relationship between the degree or severity of menstruation and the breeding system of mammals. More promiscuous animals should be taking in more bacterial-laden sperm and, thus, would have a greater need to flush everything out, or so the argument goes.

Beverly Strassmann, she of the 30-plus years of field work with the women of Mali, wrote an article about the origins of menstruation a few years later. In her paper, she refuted, point by point, each of the predictions born of the anti-pathogen hypothesis. She cited studies that demonstrate the uterine environment is, in fact, more laden with pathogens *after* menstruation than before. She argued this makes perfect sense because "blood contains iron, amino acids, proteins, and sugars, and therefore is an excellent culture medium for bacteria."[14]

The link between promiscuity and menstruation also did not hold up upon closer inspection. Primates with more promiscuous mating systems do not, in fact, menstruate more copiously than those with less promiscuous mating systems. In addition, traditionally, menstruation would have occurred relatively infrequently, as women were typically either pregnant or nursing. Sex still occurs during those stages of life. Therefore, it seems unlikely menstruation evolved as a defense against pathogens transported by sperm given that women cannot use the defense while carrying and nursing children.

These arguments were a death knell for the anti-pathogen hypothesis. As it always does, science kept plodding forward, and the refuted hypothesis did at least usher in a new wave of interest in the subject. One new idea, put forth by Beverly Strassmann herself in the same paper in which she thoroughly undressed the anti-pathogen hypothesis, was all about energy economy. The endometrium thickens in preparation for pregnancy. If pregnancy does not occur, Strassmann argued that the lining breaks down because it is too energetically costly to maintain it in the thickened state. She referenced data showing the massive increase in oxygen consumption by the thickened endometrium compared with the amount used in its regressed state. Most animals break down the built-up layer and reabsorb its various elements (this is covert menstruation). In some animals the layer is too thick to reabsorb, and the only option is to eliminate the excess blood and tissue (this is overt menstruation).

The energy economy argument contends that, since for most mammals there is a relatively short window during which females are fertile, it does not make sense to keep the endometrium thick at all times. It would be a different story if animals were fertile around the clock. Then it would be justifiable, energetically, to keep the endometrium in a permanently thickened state. To put it another way, there is no sense in always having the lights on if there are only a couple of nights when someone might stop by. It makes more sense, energetically, to turn the lights off, put on some sweatpants, and curl up with a good book (about the evolutionary origins of aches and pains) until the fertile window returns.

YOU'LL FEEL JUST A SCRATCH

This idea of Strassmann's moved the conversation much closer to the root causes of menstruation, but it still did not fill in all the puzzle pieces. It did not fully explain why some species must slough their endometria while others can reabsorb the layer. It's tempting to think the potential size of the fetus is the driving factor. Comparing various mammals, however, clearly demonstrates that size does not dictate the behavior of overt menstruation. Giant mammals like elephants and hippos practice covert menstruation, whereas the diminutive spiny mouse and elephant shrews join humans in having overt menstruation.

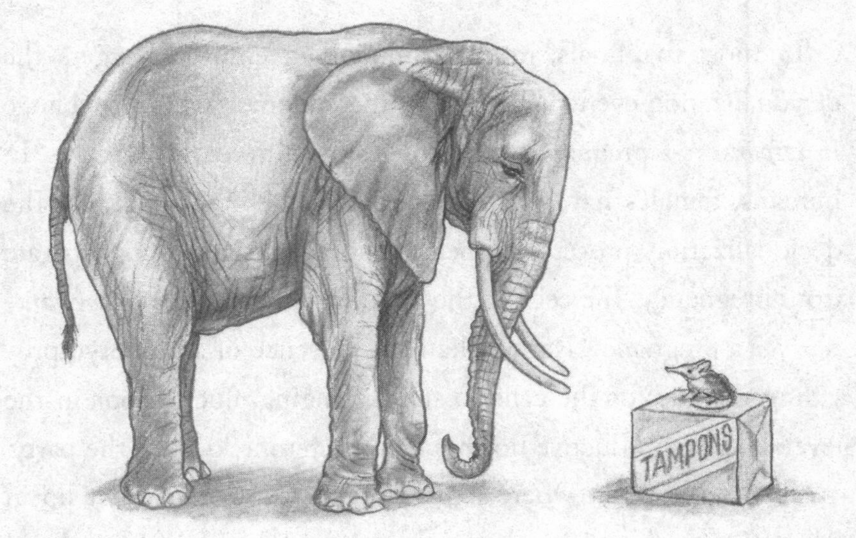

Like most other mammals, elephants reabsorb their menstrual blood and tissues. A much smaller fraction of mammals, including elephant shrews and humans, discharge them instead.

A little more research on the elephant shrews and their menstruating brethren helped solve this mystery. The important clue was learning which species engage in a physiological phenomenon called spontaneous decidualization. Spontaneous decidualization is confusing and a bit of a mouthful. Decidualization refers to a change in the cells of the uterine lining. The shape and structure of those endometrial cells can morph, and the changed cells can produce different molecules that facilitate the nurturing of an embryo. Upon changing, the uterine layer is known as the decidua. The word *decidua* originated because animals can shed the decidua layer, like leaves shedding from deciduous trees in the fall. The changing tissues recruit immune cells, and new blood vessels are generated.[15]

In most mammals, implantation of an embryo triggers the decidualization events. The cells of the endometrium thus change *in response to* a pregnancy. The situation is different in humans. In humans, females have hormonal control over their wombs. The decidualization process happens spontaneously and is separate from pregnancy. The cells of the endometrium change *in preparation for* a pregnancy. Rather than the presence of an embryo providing the cue for the cells to start changing, fluctuations in the level of the reproductive hormone progesterone kick off the party.

This is the point where it paid off for someone to get up in the business of elephant shrews. Not only did the elephant shrew hoo-ha researchers determine that those little beasts menstruate, they also learned elephant shrews undergo spontaneous decidualization. In fact, the behavior tracks very cleanly with menstruation. Mammals who experience spontaneous decidualization menstruate. Mammals with fetus-induced decidualization do not.

In order not to fall into the "pirates cause global warming" correlation versus causation trap, we need to determine if there is something more than this correlation between menstruation and SD. I wanted to hammer home the term spontaneous decidualization, so I waited to bust out SD as its abbreviation. I think you're ready, so from now on SD stands for spontaneous decidualization.

The way to move beyond correlations is to run well-controlled experiments rather than just look for relationships between two variables. Researchers have conducted those experiments investigating the possible connection between SD and menstruation. They took mice, who, under normal conditions, do not experience

SD or menstruation and gave small scratches to their uterine linings to simulate pregnancy.[16] Tiny scratches were enough to trick the endometrial cells into decidualizing. Their little mouse bodies thought they had become pregnant and changed the cells of their endometria. The researchers then experimentally blocked the receptors for circulating progesterone, and the mice menstruated. The lack of active progesterone signaled that, in fact, pregnancy had not occurred. Normally, the mice would have just reabsorbed their built-up uterine linings in the absence of pregnancy. The experimental mice had changed their uterine linings in response to the scratching of their uteruses, and thus, their endometria acted as if they had undergone SD. The mice shed their linings instead of reabsorbing them. The control mice in the experiment did not receive the endometrial scratch (so their cells did not decidualize). When those mice had their progesterone experimentally blocked, they reabsorbed their built-up linings and did not menstruate.

MIXING THE CAKE BATTER

Confused? You should be confused if you're paying attention. Reproductive anatomy and physiology is inherently complex. It arrives at the end of the year-long sequence of anatomy and physiology courses so students can have the best possible handle on basic cell biology and endocrine physiology before attempting to understand reproductive cycles. I tell students if they are not at least a little confused, particularly with female reproductive physiology, then they are not trying hard enough. The first step to understanding a difficult topic is accepting and leaning into some level of confusion.

With my reader-confusion spidey sense tingling strongly, it means it is time for another metaphor. I think a good way to explain SD and its role in menstruation is to think about baking a cake for friends who might stop by for dessert. It's not guaranteed they will visit. Maybe they have a babysitter who often cancels. But you want to be ready to whip up Grandma's famous lemon poke cake if they come over. You have everything you need laid out on the counter (flour, baking powder, sugar, butter, eggs, milk, vanilla extract, and lemons) and will spring into action the minute they arrive. To make the analogy tighter, pretend, also, that you, the baker, do not like lemon cake. You are making it only because your guests like it. At this point, there are two possible outcomes. If the guests show up, then you whip it all together into batter and put it in the oven. It is no problem if the guests bail on you because you can simply put everything back in the refrigerator or pantry. Maybe you make some lemonade with the lemons.

This first scenario matches how most mammals roll with the decidualization of their endometria. The batter is mixed (the tissues of the endometrium undergo differentiation) only if the guests arrive (if implantation occurs). If implantation does not occur, the endometrium still has not changed dramatically and the ingredients are saved instead of wasted. Because you didn't crack the eggs or mix in the vanilla extract, you can use them for other purposes.

Menstruating species are different because they mix up the cake batter *before* the guests arrive. They experience SD, in which their endometria change before implantation occurs. They get ready for the party before it is even clear if there will be a party. Menstruating species crack the eggs and cream the butter before they get pregnant. It's like they know the guests will be starving and want to have the cake ready to pop in the oven the minute the doorbell rings.

In this second scenario, there are still only two possible outcomes, but one of them is markedly different than before. The first outcome is basically the same. The guests show up and you pop the batter in the oven. Thirty minutes later and your friends are tucking into lemon cake. The second outcome is very different. In this case, if your friends' babysitter bails on them and they can't make it, you're out of luck because you've already mixed the batter. You can't get eggs back in their shells, and you can't uncream the butter and the sugar. Since you think lemon cake is gross, the only option is to throw it all away.

This is what happens most of the time in women. Their bodies mix the batter beforehand, and the guests typically don't show up. The endometrium undergoes change before implantation (this *is* SD), most of the time implantation does not occur, and there is no choice but to discard the cake batter. Just like you can't keep cake batter on the counter for a month, it is not an option to keep the endometrium permanently in its decidualized state. The only way it remains decidualized is if progesterone levels remain high. The only way progesterone levels remain elevated is if implantation occurs. In the absence of implantation, the thickened endometrium is either reabsorbed or shed. If the ingredients have already been mixed, if decidualization has already occurred, shedding is the only option. I like this metaphor, but it is going to make it difficult to look at Grandma's lemon poke cake the same way ever again.

Pregnancy-induced decidualization **Spontaneous decidualization**

In pregnancy-induced decidualization, the ingredients can be put away and saved for another day if they are not used. In spontaneous decidualization, the batter must be thrown out if the cake is not baked.

WHY MIX THE BATTER?

So it turns out the million-dollar question is not "why do women menstruate?" Women menstruate because they experience SD. The million-dollar question is "why do women experience SD?" Why do they mix up the cake batter before the guests arrive?

Given the rare and disparate nature of the behavior, it is most likely SD was not present in the ancestors of placental mammals. Instead, SD and menstruation likely evolved separately in select mammalian groups. When evolution converges on a trait independently in multiple lineages, it usually indicates there is some type of benefit involved. One would certainly hope there is some benefit to offset the cost of losing those gallons and gallons of blood over the course of a lifetime.

There are two leading ideas for the benefit an animal derives from SD. Both of them are born of a concept deemed the maternal–fetal conflict hypothesis. We like to think of the bond between a mother and her unborn child in warm and fuzzy terms. In this view, everything is set to the soundtrack of a music box, with images of fluffy clouds, rainbows, and butterflies. Pregnant women spend blissful days in rapt anticipation of the incredible gift of life soon to come. I suppose there are some women (probably, again, the annoying ones who menstruate, like, a teaspoon over the course of their 36-hour periods) for whom the peaceful, fluffy-cloud imagery of pregnancy is accurate. They just glow through the entire nine months and seem utterly unfazed and possibly even enhanced by the whole scene.

The maternal–fetal conflict hypothesis suggests the reality of pregnancy is not so idyllic. A more accurate soundtrack would be grungy heavy metal instead of a delicately melodious lullaby. Instead of glow and bliss there is, more typically, barfing, overwhelming fatigue, and conditions like gestational diabetes. The nature of maternal–fetal conflict drives to the distinctly different genetic interests of the two parties. Because the mother and child do not have identical sets of DNA, their biological motivations are different. The mother will nurture the developing fetus, but within limits. Her body will provide for the fetus, but not so much that it affects her ability to successfully carry other children in the future. The fetus really only cares about itself. It needs to keep the mother alive but is going to extract every last nutrient it can from her in the process.

How this conflict plays out can vary tremendously by taxa. Horse and pig fetuses, for example, do not burrow very aggressively into the womb. The membranes surrounding the fetus are several layers of tissue removed from the maternal blood supply. There is still maternal–fetal conflict in those species, but not the same degree of conflict seen in species where the fetus digs in further. Dogs and cats are somewhere in the middle. Their fetal tissues start to invade the maternal tissues but are still distanced somewhat from the maternal blood vessels. In the most aggressive version of placentation, the fetus roots in, like a mole into dirt, and snuggles right up against the blood vessels of the mother.

You can probably guess which type of fetuses humans ended up with. We got the uberaggressive model. And again, we see the comparative approach pay off in solving this riddle of why SD evolved. The animals that exhibit SD and menstruation are also the ones with the most invasive fetuses. Some scientists think SD evolved as a preemptive degree of protection against a hyperinvasive fetus. The logic goes that a woman gets ahead of the game and builds in some extra protection before the vampire-fetus arrives so that her unborn child does not completely suck her dry. After all, if you know a vampire is coming to your quaint, remote village, it makes sense to start beefing up the defenses of the village before the little bloodsucker gets there. Get the garlic planted, the stakes sharpened, and the mirrors shined in preparation.

There is evidence to support the idea that SD defends the mother's tissues. Although SD does ultimately change the endometrium in a way to facilitate pregnancy, several of the features of the decidualization process are aimed specifically at keeping the fetus from burrowing too deeply. The cells of the endometrium bind themselves together in a tighter manner during decidualization. An article about the evolution of menstruation published in the journal *BioEssays* discusses how the endometrial cells also produce compounds that "inhibit the activity of invasive fetal proteins."[17] In addition, the uterus ramps up the activity level of a specific type of cell called a natural killer (NK) cell. NK cells are a type of white blood cell typically used in the body to fight off tumors or cells infected with viruses. They attach to the foreign entity and release chemicals that induce apoptosis (programmed cell death) within the invader. Except this time, instead of a cancer or a virus, the "invader" is a growing unborn child. The NK cells help limit the degree of invasion of the aggressively burrowing fetus. All this activity indicates a woman's body is defending the village rather than baking cookies and setting out clean towels and sheets for the impending guest.

It is perhaps easier to understand the importance of such built-in prepregnancy preparations if we focus on a scenario in which they do not happen. A research group compared the apoptosis of embryonic cells in uterine pregnancies and tubal pregnancies (where the embryo implants within the Fallopian tube as opposed to the uterus). Tubal implantation occurs in around 2% of pregnancies and can be very dangerous if not treated. They published their results in the journal *Placenta*, which takes the title from *The*

Knee as the most fantastically specific journal name referenced in this book. The group found that the apoptosis-inducing reactions that limit fetal invasion in the uterus fail to take place in the Fallopian tube.[18] A fetus in a Fallopian tube, apparently unaware it ended up in the wrong location, puts its nose down and burrows away, entirely unimpeded, potentially even leading to rupture of the tube.

As it turns out, those built-in prepregnancy preparations are rather important. Without SD limiting fetal invasion, the vampire would run pell-mell through the village, not stopping until it quenched its thirst and put the mother on life support. The benefit of the SD protection plan is that it limits fetal invasion to a point that makes pregnancy at least manageable. The cost is that you cannot have the plan without also signing up for decades of menstruation.

EMBRYO BOOT CAMP

The second idea for why SD evolved relies on a growing body of evidence suggesting SD allows a woman's body to screen embryos. Having a built-in screening process would allow women to invest the considerable time and energy of pregnancy in high-quality offspring only. Back we go to the other menstruating species to understand this hypothesis. We have more in common with the elephant shrews and spiny mice than just SD. Another behavior linking animals (including us) who experience SD, menstruation, and hyperaggressive placentation is extended copulation. Extended copulation does not mean what it sounds like. The phrase does not refer to humans or elephant shrews going at it for hours on end.

Animals with extended copulation have sex independent of ovulation. Most species of mammals do not engage in extended copulation, and females are receptive to mating only around the time of ovulation. A female comes into estrus (or "heat," as it's commonly called), takes a mate, and then kisses off sex until the next time she comes into heat. Most mammals operate this way, from dogs, to camels, to walruses. Animals with estrous cycles vary in how often this period of female receptivity comes around. In polyestrous mammals (like cows and pigs), the cycles occur throughout the year. Seasonally polyestrous mammals come into heat multiple times a year, but only during certain seasons. This is the setup for mammals like deer, goats, and cats. A doe will come into estrus in a period referred to as the rut during the fall. If she becomes pregnant after her first cycle of the rut, then she will not come into estrus again until the next year. If she fails to become pregnant, then she will cycle again in the same rut. The last type of mammals with estrous cycles are monoestrous, in which the females are receptive only once a year. This is the pattern for dogs and bears, which I guess helps explain why male bears always seem so grumpy.

In contrast, females of animal species that practice extended copulation are generally receptive throughout the entire reproductive cycle, not just around ovulation. The ramifications of extended copulation are significant. Because mating occurs throughout the cycle, sometimes sperm remain in the female reproductive tract for days before encountering an egg. Human sperm can live in the uterus and Fallopian tubes for the better part of a week. Most die within a day or two, but some are still kicking after five or six days. How much of an impact this has on the health of an embryo is a debatable point, but there is general agreement that fresh sperm are the best sperm for fertilization, and some researchers have suggested that extended copulation may lead to a greater number of impaired embryos.[19]

A woman will make no bigger investment in her life than the one she makes in a child. There is evidence to suggest SD allows a woman's body to screen for high-quality embryos and, consequently, prevent investment in low-quality embryos. As we'll see in detail in the next chapter, this screening process is important because, for reasons not entirely well understood, a shockingly high number of human embryos have chromosomal abnormalities.

Even historically, the process of carrying, birthing, and rearing a child would have involved a huge sacrifice of time and energy for a mother. Our ancestors would have greatly benefited from any type of mutation allowing for investment in only the healthiest embryos. It would have meant time and energy saved for ancestral women. They could have put that saved time into rearing other children. These days becoming a mother includes pregnancy, birth, nursing, diapers, day care, tweenage drama, teenage drama, paying for college, and having a 24-year-old live in the basement (rent-free) while they work as a barista. It is one egg in one very important basket.

The decidualized cells of the endometrium are able to detect chemicals secreted by abnormal human embryos.[20] The cells need to be decidualized in order to detect any degree of abnormality developing. Detection of abnormality initiates a natural process that can lead to early miscarriage. This is not an uncommon occurrence in humans. Some 30% of successfully implanted embryos never make it to six weeks of gestation.[21]

The embryo selection process needs to take place early, which is exactly why it benefits women to have the cells of their endometria decidualize spontaneously, rather than waiting for an embryo to arrive. The selection committee is in place and ready to go before implantation occurs. If embryos are not vetted because SD does not fully or properly occur, any old embryo can start digging in. If this happens, an abnormal embryo can end up implanting and growing and may lead to a miscarriage weeks or even months later. There is variation in the efficacy of this embryo selection process, and women with impaired decidualization reactions are far more likely to end up having recurrent pregnancy loss.[22]*

Yet again, we see the importance and value of SD. It not only keeps fetuses from sucking mothers dry but also allows for natural investment in the healthiest possible pregnancies. Even if it does cause the royal pain that is menstruation, at least there are some benefits to balance the cramping and blood loss.

---

*A pregnancy is called a "biochemical pregnancy" if it can be detected only via urine or blood hormone testing. It is considered a "clinical pregnancy" if it can be detected via ultrasound (which usually happens around five to six weeks of gestation). Recurrent pregnancy loss is the term used if a woman has two or more clinical pregnancy losses (miscarriages) before the pregnancies reach 20 weeks.

THE COST OF CONTROL

Why do women menstruate? Women menstruate because their bodies have more control over their reproductive cycles than most other mammals. The evolution of SD allowed for that control. Once SD evolved, menstruation was inevitable. More control was ultimately beneficial because of the unique nature of human pregnancy. Maybe it was necessary to defend against the aggressively burrowing human fetus. It might have been beneficial because human embryos are uniquely error-prone, and the evolution of SD gave women's bodies a way of making sure they committed only to the healthiest pregnancies. It may have been both of those reasons or yet another reason scientists will discover with more research. It is important to remember that the evolutionary underpinnings of menstruation were largely a mystery until recently; only in the last generation have we started to get a solid handle on why it happens. There will undoubtedly be more revelations in the years to come.

I'm also confident that having a better understanding of the process won't make menstruation any more comfortable. But if you're a woman reading this, maybe it will help to know you are not alone in the animal kingdom. Somewhere out there in the wild there is a baboon, an elephant shrew, and a leaf-nosed bat having their period. And there are definitely no tampons for leaf-nosed bats, so you've got that going for you.

8

Absence Makes the Heart Grow Fonder . . . and the Penis Thrust More Deeply

How long is the average erect gorilla penis?

a. 4 centimeters (1.6 inches)

b. 13 centimeters (5.1 inches)

c. 22 centimeters (8.7 inches)

d. 31 centimeters (12.2 inches)

It is time to get down to the business of making babies. It is a fun business to be in, but it can also be stressful and messy. There are just so many fluids. The fertility of couples falls into three categories. In the first category (the "fertile as bunnies" group), we have the couples who seem to conceive with nothing more than a glance. It's almost like they just hold hands at the movies and get pregnant.

Everyone knows one of these fertile-bunny couples who reproduce with very little difficulty. You likely even know someone who got pregnant who was taking specific measures to *not* get pregnant.

For couples in a middle category of fertility, the process is not quite as effortless, but a few months of planning ("Honey, it's *time*") usually does the trick. These couples are not exactly popping out kids like bunnies, but they also do not have to upend their lives to have children.

Then there are those in a third category who try and try again yet are unable to conceive. The pain of infertility is a different kind of pain from most others covered to this point. It is a piercing psychological pain that builds with every passing month without a positive pregnancy test. To couples in this category, the entire world seems able to conceive except for them, and there is no escape from all the happy families with their happy babies and the happy mothers pregnant with their future happy children.

Infertility strikes each gender roughly equally. In one-third of infertile couples, the anatomy or physiology of the man is the cause of infertility. One-third of the time the woman is infertile, and one-third of the time the root causes cannot be identified or both partners have fertility complications. The generally agreed-upon definition of infertility reserves the word for couples who are unable to become pregnant after one year of regular unprotected sex. Somewhere between 10 and 15% of couples meet that definition.[1]

Infertility is not a phenomenon born of modern times. There are present-day factors to address, but studies going back to the

19th century show infertility was a problem even before the world was overrun with pollution and french fries.[2]

Why are some couples unable to bear children when bearing children seems like it should be a fundamentally natural thing to be able to do? Why are bunnies able to breed like bunnies, while for so many people the only option is to turn the entire process into a long, drawn-out science experiment?

FERTILITY PROBLEM #1: ENDOMETRIOSIS

To start the discussion, we briefly return to the topic of the last chapter. It is possible for some of the menstrual blood and tissue to take a wrong turn and go up a Fallopian tube instead of passing out through the vagina. This backward flow is called retrograde menstruation. The blood runs up the tubes, and because there is a small space between the end of each tube and each ovary, it exits into the pelvic cavity. No one seems to know exactly why this happens, but it appears to occur to some degree in most women. Instead of being voided, the sloughed endometrial cells can remain embedded in the body, outside of the uterus.

The migrant cells don't forget where they came from. Upon the next menstrual cycle, the extrauterine cells may continue to behave like uterine cells. This means they may build up and then break down as part of the next menstrual cycle. This time, however, there is zero chance the extra tissues and blood will exit through the vagina, as the rogue endometrial tissues are no longer anywhere

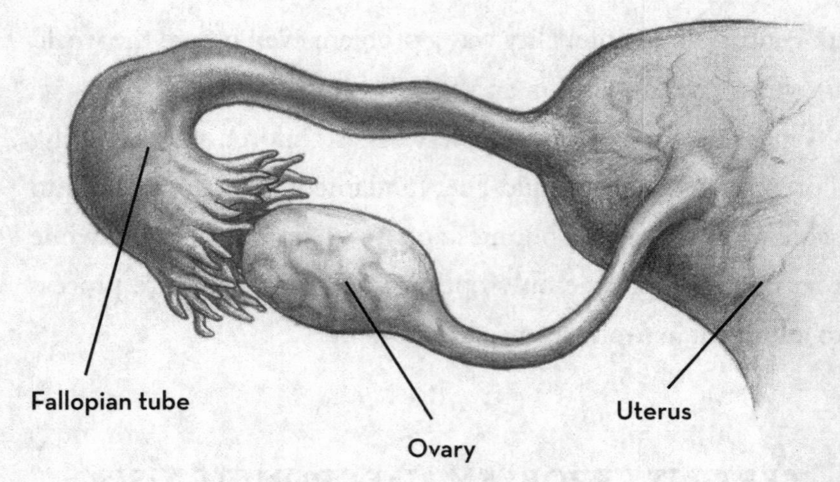

The small gap between the ovary and the Fallopian tube allows retrograde menstrual flow to exit into the pelvic cavity.

near the vagina. Each cycle exacerbates the problem of wayward endometrial cells as more and more cells break free and continue to find their way to far-flung locations. In extreme cases, doctors have found extrauterine tissues in nearly every conceivable location in the female human body, including the gut, heart, and lungs.[3]

The buildup of uterine tissue outside the uterus and the suite of symptoms it causes are referred to as endometriosis. One of the primary symptoms of endometriosis is debilitating pain during menstruation. Retrograde menstruation is not the entire reason behind endometriosis. Even though most women likely experience some degree of retrograde menstruation, endometriosis occurs in only around 10% of women. Those numbers suggest it takes more than retrograde menstruation to push a woman's body over the endometriosis cliff. The endometriosis puzzle still has many missing pieces.

There are, without question, many other factors at play involving a combination of genetic and environmental influences that make some women more susceptible to the problem than others.

In addition to physical pain and suffering, there is a clear link between endometriosis and infertility. Up to half of infertile women report having symptoms of endometriosis. It is a highly inflammatory condition, and the inflammation, combined with the remodeling of reproductive tissue, can lead to scarring and disruption, preventing the joining of egg and sperm and inhibiting successful implantation.

We know surprisingly little about endometriosis, but its link with infertility means there will be no shortage of research dollars thrown at the topic. Unfortunately, as we're about to see, it is also just one of many possible fertility hurdles that couples must clear on the path to successful conception, implantation, and pregnancy.

FERTILITY PROBLEM #2: CONCEALED OVULATION

Recall from the last chapter that one of the unusual characteristics of human reproduction is extended copulation. Females do not take mates only around their fertile windows, but instead have sex throughout their reproductive cycles. A fair amount of physiological deception goes part and parcel with extended copulation. One critical element in the evolution of extended copulation was the concealment of ovulation. Females could garner the benefits of extended copulation only if males were kept mostly in the dark about when females were most fertile.

The ovulation of most mammals is anything but concealed. Female cats make such a racket about the whole thing it can wake up an entire neighborhood. In addition to pheromonal signals, female dogs exhibit some obvious visual cues, including a swollen vulva and bloody discharge, that really get the hounds worked up. Signs of estrus in many primates follow a similar pattern, with engorged genitalia around the time of ovulation being the most obvious cue.

Humans have, albeit somewhat debatably, concealed ovulation. Most couples do not have a proof-positive way of knowing exactly when ovulation occurs. Some women (about 20%) do experience pain, ranging from minor to intense, around the time of ovulation, referred to as *mittelschmerz* (from the German meaning "middle pain" because it occurs around the midpoint of the cycle). Doctors think ovulatory pain is due to either the stretching of the ovary prior to release of the egg or irritation of the abdominal wall from the fluid discharged by the ruptured follicle (the group of cells that surround the egg prior to ovulation). There are often other physical signs of ovulation, including changes to the viscosity of the cervical mucus, but nonetheless many women (and their partners) are left largely in the dark about the timing of peak fertility.

Concealed ovulation is thought to have evolved in females because it provides concrete benefits. To begin with, the behavior may promote male fidelity. If a male cannot tell when a female is fertile, then he has to hang around all the time to ensure he has access to her when she *is* ovulating. Otherwise, the male might leave at the wrong time and a rival male could sneak in there during her fertile window.

By concealing ovulation and mating throughout the reproductive cycle, females may have hit upon a way to increase the investment on the part of their male partners. Most male mammals peace out immediately after mating season, but human males tend to stick around. Concealed ovulation and extended copulation of females likely play an influential role in motivating males to stay close. Back in the day, increased male fidelity would have translated into more resources and greater protection for females and their offspring. In addition to helping with resource accumulation in modern times, male fidelity also means there is a partner to help change diapers or drop the kids off at school. Helping females with child-rearing and chores is not a behavior found in most male mammals. Male honey badgers are not exactly waking up in the middle of the night to give the baby a bottle.

This first explanation for the evolution of extended copulation and concealed ovulation is nice and warm and fuzzy. By promoting monogamy, extended copulation and concealed ovulation made it so a couple would raise a beautiful family together. The second hypothesis is violent and salacious. Some have argued that concealed ovulation evolved because it reduces the likelihood of infanticide by males. Infanticide is a widespread behavior in mammals. It is perhaps most well documented in lions, but scientists have witnessed the behavior in diverse branches of the mammalian tree, from canids to felids to squirrels and ungulates. It is particularly common in primates. A review of the behavior in the esteemed journal *Science* demonstrated it is most frequently seen in social groups of animals in which "reproduction is monopolized by a minority of males."[4] When a male is replaced at the top of the pecking order, the newly dominant male will commonly kill any unweaned offspring. Upon the death of their offspring, females cease nursing and their reproductive cycles begin again. As brutal as it is, infanticide is an evolutionarily stable strategy in some species because recently ascended males are able to father children sooner with the females of the group.

Extended copulation and concealed ovulation may be a counterstrategy to infanticide. By mating outside of the fertile window and taking multiple mates, a female is able to considerably muddy the paternity waters. Males are less likely to kill offspring who might be their own, and thus, a newly dominant male is less likely to kill the children of a female with whom he mated previously.[5] Polyandrous females (those who take multiple mates) reduce the risk of infanticide because no one in the group knows for certain who the father of the children is.

Promoting male fidelity and preventing infanticide would both have been a win for early human females. These days, however, the behavior of concealed ovulation can be terrifically frustrating for couples trying to conceive. There are ways to know when a woman is fertile, but peeing on a test strip to measure the levels of luteinizing hormone (LH, a reproductive hormone that spikes prior to ovulation) is a far less straightforward process than what other animals experience. Couples with fertility issues spend copious amounts of time, energy, and resources simply figuring out the perfect time to have sex to give them the best shot at conception. This is not something other animals have to consider. In most mammals, the female comes into heat and everyone knows it is time to have sex. Bing, bang, boom, and a little while later there are babies running around.

COVER THE PUBIS

The degree to which human females conceal estrus remains a matter of some debate. The old school of thought is that the female body hides the process. Recent studies, however, complicate the story. Multiple investigators have uncovered independent evidence suggesting female physiology and behavior may change around ovulation and that males may be, at least subconsciously, aware of when women are most fertile.

What's the best way to get down to the bottom of whether females are giving off perceptible cues around the time of ovulation, signaling their reproductive status to potential mates? Obviously, you ask exotic dancers to take part in a study tracking their tip earnings by shift and then look for a relationship between those earnings and the dancer's reproductive cycles. At least it was obvious to Geoffrey Miller, a professor of psychology at the University of New Mexico.

It is rare to find research with both style and substance. The title of Miller's paper is "Ovulatory Cycle Effects on Tip Earnings by

Lap Dancers: Economic Evidence for Human Estrus?"[6] A scientific paper with the words "lap dancers" and "ovulatory cycle" in the title is clearly going to grab some attention. Miller's research goes beyond generating headlines, however, and balances style with creative, sound, well-reported, and fascinating substance.

One quick way to determine the impact of any given research is to look at the number of times other authors cited the work in the years after its publication. If a decade has passed since the publication of an article and only three other research teams have referenced the work in their subsequent publications, it indicates the research was largely ignored by the scientific community. Miller's "lap dancer ovulation" paper has been referenced more than 400 times since its publication. These hundreds of references are not by TV shows, magazines, and newspapers; they are references in peer-reviewed journal articles written by other scientists working on related questions about fertility, sexual selection, and evolution. Many papers are not referenced even 50 times, and some are referenced so few times one can assume they were read only by members of the team who did the research and maybe their mothers.

One of the first observations I made about Miller's paper was the inclusion of a section called *background*. Most research articles go straight from the introduction to the methods. Miller and his coauthors made the unusual move of sneaking in a background section. The first line of the background is "Because academics may be unfamiliar with the gentlemen's club subculture, some background may be helpful to understand why this is an ideal setting for investigating real-world attractiveness of human female estrus." That feels like a gross generalization of academics, but it is also probably accurate. I think it is safe to say most professors are not hitting up strip clubs on a regular basis. I mean, it's awkward enough to bump into students just going out to eat or when running to the grocery store.

The rest of the background section uses notably formal and scientific language to describe the ins and outs of the strip club scene in Albuquerque. It includes very no-nonsense descriptions of the practice of lap dancing ("rhythmic contact between the female pelvis and the clothed male penis") and the work-time clothing selections of the women, including their legal obligation to at all times have "garments to cover the pubis." Needless to say, this is not the usual type of description you find in the methods section of a paper, which is why the authors created a special section.

The study included a mix of women who were on birth control and others who were cycling naturally. One valid critique of the research is the sample size—there were only 18 participants. For 60 days, the 18 women tracked their tips across their shifts and also recorded the precise days they began and ceased menstruation. The study covered 296 shifts that included approximately 5,300 lap dances. That's a lot of "rhythmic contact between the female pelvis and the clothed male penis."

The researchers broke the cycle phase data into three categories. Day 1 marked the onset of menstruation, and days 1 to 5 were categorized as the menstrual phase. Days 9 to 15 were considered the fertile days for the dancers, and days 18 to 28 were called the luteal phase, representing the time after peak fertility but before menstruation began again. The study did not include days 6 to 8 in the analysis, as those days mark the transition between menstruation and fertility. The same was true of days 16 and 17, which could have fallen in either the fertile or luteal phase.

The typical dancer's shift was five hours. For women cycling naturally (meaning they were not on birth control), tips dipped around the time of menstruation and spiked tremendously around ovulation. Those women made, on average, $185 per shift during menstruation, $335 per shift during the fertile phase, and $260 per shift during the luteal phase. Women on the pill did not experience the same income spike during their "fertile" phase. In fact, women on birth control earned $83 less per shift on average than women cycling naturally! Somehow, the strip club patrons were influenced by the reproductive phases of the dancers. It would be interesting to know if men were able to detect the differences only because the women were literally on their laps. In other words, do the tip earnings of female servers or female Uber drivers also vary based on where women are in their cycles? Also, somebody needs to flip the script with this study and reverse the roles. They should have women go to strip clubs where men are dancing to see if women's reproductive phases correlate with how much they tip. There are still many unanswered questions about basic human reproductive biology. Get busy, grad students! Get busy!

HOO-AH—VERBAL FLUENCY OF A WOMAN

The results of Miller's study raise the question about what kind of signal fertile women might be sending. One possibility is that women produce different pheromones during the fertile phase, and other research suggests that the scent of a woman is more attractive to men around the time of ovulation.[7, 8] A common method for such studies is to have women wear T-shirts and then, after a period of time, have the scent of the T-shirts evaluated by men. Some have gone further than the standard dirty T-shirt tests to demonstrate that men exposed to the scent of an ovulating woman have higher levels of testosterone than men exposed to the scent of a woman who is not ovulating.[9]

There are, however, other possibilities for cue or signal transmission in addition to scent. I can't help but wonder if the women were perhaps dancing more . . . enthusiastically . . . during the fertile days of their cycles. I guess the researchers would have had to film each interaction to get to the bottom of that question, which really would make for an IRB nightmare.* Investigators have found links between reproductive phase and other cues such as waist-to-hip ratio[10] and facial attractiveness.[11] One of the most intriguing revelations is the association scientists discovered between where a woman is in her reproductive cycle and her faculty with

*IRB stands for institutional review board and is the committee at an institution that makes sure all research performed on human subjects is done on the up and up. I can only imagine they had a field day with this strip club study.

language.[12] When given 60 seconds to name as many words as they could starting with a given letter, women who were around their ovulatory window could name more words than they were able to in the period outside of their fertile time. As it turns out, the fertile lap dancers may have been manipulating the men as much with their dirty sweet nothings as with their odors. (My wife, Julie, consistently crushes me at games like Scrabble and Boggle, so I might have to put this knowledge to use and agree to play her only when she is as far as possible from her peak fertility. I'm sure I will still lose, but maybe I can make the outcome a little more interesting.)

The clothing choices made by women may also be influenced by where they are in their menstrual cycles. This would have been a difficult variable to test in the lap dancer study (not a lot of clothing there to analyze), but other research has suggested women do try to increase their attractiveness (through grooming and clothing choices) around their most fertile times.[13] If Julie comes out with her hair all done up and wearing something slinky, the obvious move is to pivot away from Boggle and toward something more romantic.

FRESH CHEESE

The issues of endometriosis and concealed ovulation are firmly rooted in our past. They create obstacles that couples must clear on the path to a successful pregnancy. If women are lucky, they are able to clear the endometriosis hurdle. Clearing the second hurdle requires a couple to successfully sync their activities and have sex at a time when the woman is ovulating. Even if the lap dancer study makes it clear human females do not entirely conceal estrus, it is also not as if women and their bodies are out there flying a giant flag announcing their peak fertility. Getting the timing right is not a trivial issue for some couples.

Assuming couples clear both of those first two hurdles, there are still several more they have to leap to end up pregnant with a healthy kiddo on the way. The focus now shifts to men and their contributions. If we view the entire process of pregnancy as a party, a woman is in charge of nearly all the planning and preparation. A woman has to make the guest list, clean up beforehand, decorate, prepare nearly all the food and drinks, arrange for some entertainment, be a gracious and charming host, and then clean up after the party. A man just stops by for a few minutes. He might bring a cheap store-bought cheese tray, or something along those lines, but his contribution is minimal compared with the woman's.

The kicker is that you can't actually have the party without the cheese tray. Even though it seems like an incredibly minor contribution compared with the investment the woman makes, this is a weird party in that it only really gets going once the cheese arrives. It's like a party in Wisconsin, where it's not considered much of a party unless there is cheese. Therefore, it's pretty darn important that the dude arrives with the cheese, and that the cheese is fresh and ready to go.

The cheese is sperm. I trust you had that figured out. (Sorry if I just ruined cheese for you.) The only biologically necessary contribution the male has to make is a donation of his sperm. In many cases, he will contribute more, but sperm is the one thing he must contribute for reproduction to occur. There is no getting around this as a human. Other animals, including various species of insects, lizards, snakes, and sharks, have evolved mechanisms allowing for development of eggs into embryos without sperm. Science has yet to turn up a mammal that can pull this off. Thus, at least in mammals, you can't have a party without the cheese.

The sperm factories themselves, the testes, are an evolutionary oddity. The testes need to remain intact and functional if reproduction is to occur. Yet there they hang, exposed and dangling, ready to be kicked in a fight or snagged on a barbed-wire fence. Would it not make a hell of a lot more sense to have them tucked up inside?

It is dangerous to have the family jewels out there swinging in the breeze. When playing baseball as a kid, I often ended up playing third base, or the "hot corner," as baseball folk refer to it. Playing third comes with the very real potential to get racked when hitters rip a ball down the line. At some point in elementary school, before I started wearing a cup, I stopped a liner with my groin instead of my glove. It completely laid me out. You do not forget such an event; it sears into your hippocampus for life. These days when I play softball, I play in the outfield. I am not taking the chance at the hot corner; some other sucker can play third.

Just having the potential to get hurt in such a way is the consequence of a unique anatomical quirk of male fetal development.

Testes start life safely tucked away in the abdomen of a male in utero. Sometime around 26 weeks of gestation, however, the testes descend from inside the abdomen to their resting position in the scrotum. They go from a perfectly safe sanctuary to an exposed and vulnerable location. The ovaries, molded from the same starting materials, do not follow suit, staying nestled within the warm confines of the abdomen.

The classic explanation given for this quirky mammalian trait is that the sperm must remain cool during production and subsequent storage. By hanging out in the scrotum instead of the abdomen, the whole manufacturing process takes place in a slightly cooler environment. It's like the testes are leaning out the window of a hot and stuffy building to catch a breeze. One testicle tends to hang lower than the other, and some have suggested this is because it allows for more exposed surface area for heat dispensation.[14] If the boys are all snuggled up next to one another, it keeps the heat trapped, or so the thinking goes.

No one seems to know exactly *why* sperm production and storage need to happen in a cooler environment in mammals. External genitalia are the exception to the rule in animals. How can all other types of critters, from beetles, to blowfish, to bald eagles, manage to keep their testes tucked up inside and not have a problem with their sperm being too warm? Even a few mammals, like elephants, have internalized testes.

One idea to explain why most mammalian sperm must stay cool is the activation hypothesis.[15] Its authors suggest that sperm, somewhat chilled until the big moment, don't start swimming like mad until they hit the warm female reproductive tract. There is evidence to support this idea. Sperm become much more active when the ambient temperature rises from room temperature to body temperature. They start racing around like it's go time. Sperm remain mobile for only a few hours at elevated temperatures. By keeping them cooler outside the body, they stay inactive until the actual time when they need to move like crazy.

I have never encountered a reasonable explanation for why mammalian sperm would need a temperature differential to be effective when it is not required in other animals. Maybe it has something to do with our warm-bloodedness. Or maybe mammals have especially convoluted female reproductive tracts. Sperm may have to ramp up their activity in mammals to make the lengthy, perilous journey all the way to the waiting egg in the Fallopian tube. The long and short of it is, we still do not know *why* human sperm must stay cool, we just know *that* they must stay cool.

The authors of the activation hypothesis go so far as to suggest humans prefer to have sex at night because of the added benefit of a reduced scrotal temperature. With a greater differential at night between the temperature of the scrotum and the internal temperature of the female, sperm may theoretically be more active during nocturnal copulation than diurnal copulation. More research is needed, and I imagine it would not be too hard to track down a group of college students willing to supply some data in exchange for extra credit. I bet those UT psych students we met a few chapters back would be up for the job.

FERTILITY PROBLEM #3: UNDESCENDED TESTES

Testes do not always make their scheduled descent into the scrotum during fetal development. Having an undescended testicle is one of the most common birth defects in males. It occurs somewhere around 3 to 5% of the time in male newborns. In 10% of those cases, the condition is bilateral, with neither testicle making the trip into the scrotum by birth. Sometimes the issue resolves itself, but if a testicle has not descended on its own by about three months of age it will most likely remain internalized. Internal testes are a serious problem because if a testicle remains undescended, it becomes infertile.

Little boys with one or two undescended testicles can undergo surgery to lower the testes, but it is now understood the surgery needs to occur within the first year of life. Even though sperm production does not kick into overdrive until puberty, the important spermatogonia (the cells that will eventually divide during spermatogenesis) develop shortly after birth and need to remain cool during the dormant prepubescent years. It's no different, really, from keeping a piece of meat frozen until you want to use it. In this case the meat is a testicle, the freezer is a scrotum, and you might not want to use the piece of meat for 30 years. It cannot stay out on the counter for the first year; it needs to get in the freezer right away. Doctors moved up the age for surgery for a child with undescended testicles with each passing decade as the importance of getting the spermatogonia into the proper environment early became more clear.

The inside-to-outside development of testicles leaves a weak spot in the abdominal wall. The passageway for the nut from the gut (sorry, couldn't resist) is called the inguinal canal, and even after testicular descent, a small opening remains as a conduit between the two locations. The externalization of the testes creates a delicate anatomical feature for the duration of the life of a male. During the big swim, sperm travel via a spermatic cord that runs back up through the inguinal canal and into the abdominal cavity before picking up a variety of secretions, making a sharp turn, and heading for daylight.

The unique anatomy of the area makes it susceptible to injury. When the abdominal muscles weaken or experience stress (like when a guy tries to lift something heavy), a piece of the intestine sometimes protrudes through the inguinal canal, creating an inguinal hernia. Sometimes it is possible to massage the bulging, misplaced piece of intestine back up into the abdomen. In more serious cases, the intestinal tissue becomes "incarcerated" and slips down into the scrotum. An incarcerated inguinal hernia causes some extremely uncomfortable bulging and can become deadly serious if the rogue intestinal tissue becomes strangulated and loses its blood supply. The only option then is to undergo surgery to get everything put back into the abdominal cavity where it belongs.

Women also have inguinal canals. In their case, ligaments that help hold the uterus in place pass through the small canals on their way to where they anchor at the labia majora (the female homolog to the scrotum). The ligaments serve as an extra barrier against any possible migration of the intestines. Inguinal hernias can happen in women, but the passageway remains a much weaker point in males, making them vastly more susceptible to inguinal hernias.

Inguinal hernias are a stark reminder of a clunky, far-from-perfect feature of male development. The need for the testes to become housed in the scrotum not only leaves males uniquely injury-prone but also creates another potential fertility hurdle for any couple in which the man had a testicle, or two, that did not make the regularly scheduled descent.

ERECT GORILLAS

I like to throw an occasional curveball of a question at my anatomy students to spice up the class. When we get to the section on reproduction, one of my favorites is, "How many sperm do you think there are in the average human ejaculate?" I tell them to not worry about being wrong and to just take a shot in the dark, so to speak. Their answers are usually terrifically low. Someone will boldly come in with an answer like 5,000 or 10,000 sperm, and I'll casually make a "more" sign with my thumb. Someone else will say 100,000 and I'll give the more sign again. Then 1 million, 10 million, 100 million. The answer is 200 million. 200 million! It is, by all counts, an obscene quantity of sperm. It is more sperm in one ejaculate than the populations of the United Kingdom, France, and Italy added together! How did sperm production spiral so totally out of control?

The answer is sperm competition, and the topic of sperm competition is a big part of the human infertility story. It also means we have reached the time to talk about gorilla genitalia. Scientists will measure anything quantifiable. In this case they've put their rulers and calipers to use to obtain precise measurements of the private parts of all manner of primates. You will recall the question at the beginning of the chapter was "How long is the average erect gorilla penis?" I bet that is not something you thought you would consider today. Take a minute and think about it. Male gorillas can weigh up to 180 kilograms (400 pounds), so keep that in mind while you mull it over. Here are a human and gorilla drawn roughly to scale for your consideration, to keep you from accidentally peeking at the answer in the next line of text.

No peeking.

The answer is . . . four centimeters (1.6 inches).* For a point of reference, a baby carrot is about five centimeters long. King Kong is not even packing a baby carrot when fully aroused. He also has tiny testicles to go along with his humble penis. Across primate species, testes size and sperm counts are related to mating systems. Some species, like gorillas, get all their competition out *before* mating. Their mechanisms of sexual selection are all precopulatory, to use the lingo of evolutionary biologists. Gorillas beat their chests, bare their teeth, and fight for access to females. The biggest, baddest males monopolize an entire harem of females. Male gorillas do not need to have extensively large genitalia because their genitalia and their sperm are not in direct competition with those of other males. Unlucky for female gorillas, I guess.

It is a *very* different story with chimpanzees. Sex in chimpanzees is more of a free-for-all, anything-goes type of situation. There is rampant promiscuity by both sexes, and females take multiple mates during estrus. How much sperm a male is able to produce, and his ability to place it deep within a female, are highly selected traits in chimps. Males with larger testes are able to produce more sperm. Males with larger penises are able to get the sperm closer to where it needs to be. Males with higher sperm production end up fathering a greater proportion of the offspring. Primates are not the only group with this link between sperm production and paternity; scientists have documented the same phenomenon in a wide range of other mammals and birds.[16,17]

---

* Reports give a range between 3–6.5 cm, with most accounts coming in on the low end of that spectrum.

Comparing testes size between gorillas and chimps drives the point home. With their reproductive behaviors making conditions ripe for strong postcopulatory sexual selection, chimpanzee males are rather well endowed compared with gorillas. Despite their much smaller body size (a male chimp weighs about 25% as much as a male gorilla), chimp testes are nearly *four* times as large as gorilla testes.[18] Those large testes lead to the production of 12 times as much sperm per ejaculate compared with gorillas.[19] Chimp penises are also twice as long as gorilla penises, despite chimps' smaller stature.

Humans fall somewhere in between gorillas and chimpanzees in terms of both testes size and sperm production. Humans produce five times the number of sperm cells per ejaculate than gorillas. The results indicate sperm competition has likely played a significant role in the evolution of human reproductive anatomy and physiology. The data suggest our historical mating systems were decidedly not chaste. We were, however, probably not getting around quite as much as chimpanzees. Interestingly, we have the biggest penises of all three primate species considered here. The reason for this is unclear, but some have suggested the human phallus is shaped in such a way to remove the sperm of previous males.[20]

Tiny gorilla dicks and chimpanzee orgies aside, the important thing to remember here is that human males make unfathomable quantities of sperm—200 million sperm per ejaculate is a lot of swimmers. The historical drive to produce vast quantities of sperm is yet another situation where humans have been backed into a corner by our past in a way that can end up negatively affecting fertility in the present.

GREETING-CARD BIOLOGY

The comparison of primate mating systems and genitalia suggests that human males make so much sperm because of the possibility their sperm may have to compete with sperm from other males. It is possible to see further evidence of the influence of sperm competition on sperm production by taking gorillas and chimps out of the picture and focusing exclusively on humans. A human male adjusts the size of his, ahem, load, on the perceived possibility his mate was recently with another male.[21, 22] Let's make this idea a little more tangible with two different romantic scenarios:

Scenario #1: Tony and Wanda are a couple, and they are on vacation. On Saturday night they take a walk along the beach, have a few drinks, and end up having a roll in the hay. They spend the next few days together, swimming and reading books by the pool, but do not have sex again on Sunday, Monday, or Tuesday. Wednesday night is their last night together before heading back to their jobs. They enjoy one last tasty vacation meal, sing some karaoke, and have sex a second time before returning to real life the next day.

Scenario #2: Tony and Wanda are a couple, and they are both in the middle of a busy time at work. They plan a date for Saturday night and end up having sex. Wanda gets up early on Sunday to leave for a business trip. She is out of town until Wednesday. Tony and Wanda do not see each other during this period, and they are both so busy they only have time to exchange a few brief text messages. On Wednesday night, Wanda arrives home and Tony, bless his heart, has her favorite meal of pasta carbonara waiting for her with a glass of wine. After dinner, they celebrate their reunion by having sex before turning in for the night.

In the two scenarios, we have the same sex patterns (sex on Saturday and again on Wednesday) but under two different circumstances. In scenario #1, Wanda was by Tony's side the entire time between their romantic activities. In scenario #2, she was off on a business trip between the two events. Research on reproductive behavior suggests Tony will ejaculate more sperm on Wednesday night in scenario #2 than he would on Wednesday night in scenario #1.

Believe it or not, scientists have collected this type of data, and by data I guess I mean semen. There are other findings from this type of research. In a study first published in *Evolution and Human Behavior* and then discussed in a review in *Current Directions in Psychological Science*, when there was perceived infidelity or separation from their mates, males "thrusted the penis more deeply and more quickly into the vagina at the couple's next copulation."[23, 24] Absence makes the heart grow fonder . . . and the penis thrust more deeply. Put that on a greeting card.

FERTILITY PROBLEM #4: SPERM COUNT

Now that we understand why guys produce so much sperm, let's get down to discussing how needing copious amounts of sperm can negatively affect fertility. On the surface, it doesn't seem like an excess of sperm should create a problem. There is, however, a potential downside to males producing a veritable flood of sperm. With hundreds of millions of overzealous sperm produced, sometimes a second sperm can slip in and fertilize the egg after the first sperm, a condition known as polyspermy. Fertilization by two sperm brings a potential pregnancy to an abrupt end. Cells like sperm and eggs are haploid, having just one set of chromosomes containing the genetic instructions for the cell. All the other cells of the body are diploid, possessing two sets of chromosomes. If two sperm fertilize an egg, the resulting zygote is triploid, with three sets of chromosomes, and a triploid human zygote is not viable.

In response to the overwhelming production of sperm by males, in order to prevent polyspermy, the female body evolved in a direction that made fertilization more of a challenge. If you view the female egg as a priceless treasure inside a castle, it takes many barbarians to breach the castle gates to get one intrepid soul through to the inside. Because of the heavy defenses, a single lonely visitor arriving at the castle cannot pull off the job. But as soon as one rogue soldier is through, the castle defenses fully kick in, and the membrane of the egg hardens to prevent another sperm from sneaking in behind the first one.

With males producing gobs of sperm to compete with other males and females producing heavily defended eggs to ensure fertilization by only one sperm, the two human sexes entered an arms race many years ago. A more heavily defended castle meant males had to produce more sperm to breach the gates. As males produced more sperm, the defenses had to be even stronger to prevent polyspermy. With stronger defenses to overcome, males had to produce even more sperm. More sperm. More defense. More sperm. More defense. More sperm. More defense.

Sperm face other obstacles well before the castle gates. Hundreds of millions of sperm start the trip, but the journey has become so arduous that only a fraction of those sperm ever make it to the egg itself. Only a few thousand sperm survive the perilous trip through the vagina and the uterus, and then many of those head up the wrong Fallopian tube, where nothing is happening that month. When it's all said and done, only a few hundred sperm may find themselves in the vicinity of the precious prize. The stout defenses of the female reproductive tract not only prevent polyspermy but also ensure that only the strongest, fittest sperm have the opportunity to fertilize the egg.

It is easy to see how this back-and-forth has had a potentially negative impact on fertility. The system spiraled out of control to a point where males and females are more like reproductive combatants than collaborators. A reasonably high sperm count is necessary to increase the odds of success of fertilization. It cannot be too high, however, or fertility may be impaired by polyspermy. It also cannot be too low, or not enough sperm will reach the egg. As ridiculous as it sounds, a sperm count is low if an ejaculate contains fewer than 39 million total sperm. Sperm competition and the resultant sexual conflict and arms race between the sexes has pushed the situation to an extreme.

For many males, low sperm count, or oligospermia, becomes a significant fertility issue. Being in a committed and trusting relationship potentially compounds the issue of oligospermia. In cases where low sperm count is a problem, fertility counselors would be wise to recommend that their patients spend some time apart before having a rendezvous in bed. Remember how Tony increased his sperm count in scenario #2 when Wanda returned from her business trip? This strategy certainly would not solve everyone's fertility issues, but it might just provide the boost in sperm count necessary for some couples to get one lone knight past the walls of the castle.

FERTILITY PROBLEM #5: ERROR-PRONE EMBRYOS

We are getting closer to having a complete picture of the evolutionary underpinnings of human fertility problems. Let's assume a couple clears all the hurdles discussed thus far: no problems with endometriosis, reproductive timing, or sperm count. One sperm and one sperm only gets in there and fertilizes the egg. We should be good to go then, yes? Weirdly, no. In fact, we have not yet hit upon what many see as the most significant issue with human fertility. Up to two-thirds of fertilized human eggs fail to develop somewhere along the path between conception and birth. Some fertilized eggs never implant. In those cases, women do not even miss a single menstrual cycle. In other cases, the embryo implants but the pregnancy does not progress, and a miscarriage occurs in the ensuing weeks to months.

As introduced in the previous chapter on menstruation, one reason miscarriages happen is because humans commonly make error-prone embryos. A high percentage of human embryos have chromosomal abnormalities. If too many of the cells in a developing embryo have chromosomal mishaps, the embryo will not be viable. Aneuploid cells are cells with too few or too many chromosomes. Pack a cell with 23 pairs of chromosomes and everything should be fine. Sneak an extra chromosome in there or remove one and all bets are off. Aneuploidy often results in the natural termination of a pregnancy. Sometimes, though, the pregnancy remains viable and the baby is born with a condition like Down syndrome or Turner syndrome, in which the child has either an extra chromosome or a missing chromosome throughout the cells of their body.

The mistakes are not always as dramatic as having an entire extra or missing chromosome. Pieces of chromosomes can also swap around in ways that lead to trouble. Such translocations or imbalances can give rise to developmental errors that result in a miscarriage or, if the pregnancy is carried to term, cause significant birth defects.

Infertility is sometimes driven by preconception problems when the sperm or the egg brings an irregular number of chromosomes or misshapen chromosomes to the table. Preconception errors are more likely to happen the longer gametes sit around before an individual decides to have children. The accumulation of errors within gametes is one of the main reasons everyone's fertility (both males and females) declines with age.

It is even more likely, however, for chromosomal issues to arise postconception.[25] Perfectly normal and healthy gametes with the correct number of chromosomes routinely come together and produce embryos with aneuploid cells or other chromosomal imbalances. The condition of aneuploidy is so frequently observed in human embryos that some researchers are starting to consider the idea that it may, in fact, be the normal, natural condition for humans. A systematic meta-review of the topic revealed that 73 out of 100 human embryos contain at least some aneuploid cells.[26]

Some human embryos are able to right the ship. They overcome a few aneuploid cells, go on to have normal development, and result in the birth of fully healthy children. Other embryos end up naturally aborting because too many cells are chromosomally aberrant. In a third category, embryos develop into fetuses and make it to term, but the children are born with developmental disabilities.

Why are early human embryos so prone to chromosomal imbalances? No one knows the answer. One hypothesis suggests our system of birthing only one child at a time may play a role. In order for the mother's body to accept the pregnancy, the embryo needs to produce adequate amounts of the hormone human chorionic gonadotropin (HCG) to signal the presence of a viable embryo to the mother. Over-the-counter pregnancy tests detect the presence of HCG in urine. In animals that birth litters, all of the bunnies, kittens, or puppies can bear, together, the responsibility of making enough HCG. If researchers implant just one mouse embryo into a female mouse, she usually rejects the pregnancy. The one embryo does not make enough HCG for the mother's body to recognize and make appropriate physiological changes to support the developing offspring.[27]

Typically, in humans, a single embryo has to do all the work. The pregnancy will come to a halt unless the embryo quickly ramps up production of HCG. To crank out enough HCG, the embryo must aggressively divide within the first few days. The rapid pace of production may lead to errors. Mistakes occur whenever caution is thrown to the wind. People who drive too aggressively get into accidents. Soccer or football players who play without restraint turn the ball over. In order to simply and quickly sound the HCG horn, human embryos may divide in a way that ultimately dooms many of them to failure.

Scientists and researchers have only recently learned that having aneuploid cells as part of the embryo may, in fact, be normal. Understanding why it happens and what to do about it are the obvious next steps. The revelation calls for further comparative studies involving various mammalian subjects. One way to test the one-offspring hypothesis would be to look at fecundity rates and the degree of aneuploidy in mammals based on the sizes of their litters. Do other mammals who birth one offspring at a time tend to produce embryos as chromosomally off-kilter as human embryos? In other words, do elephants, rhinos, whales, and orangutans have to work through the same aneuploid issues as humans? If not, if their embryos are more chromosomally perfect, how do they do it? How do they manage to maintain single-offspring pregnancies without accumulating gross numbers of early chromosomal errors? Several questions remain unanswered, which is true of many aspects of human fertility.

FERTILITY PROBLEM #6: LAST BUT NOT LEAST—THE MODERN DILEMMA

Humans have a unique number of fertility issues. Many females must navigate conditions like endometriosis. Many males have low sperm counts, making it difficult to overcome the natural defenses females put up to guard against too many sperm. Our embryos are incredibly error-prone, leading to the chromosomal imbalances, deficiencies, and excesses that make up the most significant burden on human fertility.

Impacts from a modern lifestyle compound these fertility challenges. As we've seen before, thorough explanations of human shortcomings require looking at both past and present human behaviors. Our jaws are already small for our teeth, and then we double down on the problem by not chewing enough during the formative years. Our throats are susceptible to choking, and we push the issue with early exposure to grapes and hot dogs. Our knees are naturally injury-prone, and then we press our luck cutting and pivoting while playing basketball and soccer.

In the case of fertility, the doubling down is occurring in two different ways. (I guess that makes it quadrupling down.) First, people are having children much deeper into their lives than at any point in human history. For men, both sperm count and sperm motility gradually decrease with age (especially after 40). For women, the delay makes it more likely they will need to navigate additional complicating matters like uterine fibroids (muscular noncancerous tumors that grow in the walls of the uterus), which are much more common in older women and can negatively affect fertility.[28]

In addition, with many couples waiting until their late 30s or 40s to procreate, the supplies of sperm and eggs used to make children have gathered far more dust than at any time in human history. All that dust translates to more errors within the genetic and chromosomal architecture of the sperm and the eggs. It is hard enough to make a healthy embryo even if a man and a woman contribute a sperm and an egg that are genetically and chromosomally perfect. The longer sperm and eggs sit around before they are used, the more likely they are to end up acquiring some mistakes, leading to a consequential decline in fertility.

The second piece of the modern dilemma is more of a head-scratching phenomenon. The sperm counts of men in Western civilizations have declined precipitously in the last generation. Both sperm concentration and total sperm counts have dropped in North America, Europe, Australia, and New Zealand by more than 50% since the 1970s.[29] The average concentration of sperm in those populations has fallen from 99 million sperm/ml of semen to 47 million sperm/ml of semen. Forty-seven million sperm/ml still sounds like a lot, but don't forget, it takes millions of sperm to pull off the job naturally. The concentration in a sample is considered low if it is under 15 million sperm/ml. The decline in sperm production has not occurred, to date, in South America, Asia, and Africa.

With average sperm concentrations cut in half in Western countries in less than 40 years, there has been a lot of handwaving about pollutants, diets, and modern lifestyles as the potential culprits. To date, however, scientists have not successfully pinned down specific causes. If sperm production keeps declining (and to date it shows no signs of leveling off), the situation could become dire relatively soon. A worldwide fertility crisis could move from the realm of science fiction into the realm of scientific reality.

BUCKLE UP

Despite this somewhat apocalyptic outlook, plenty of women are still getting pregnant. In the face of endometriosis, undescended testicles, low sperm counts, chromosomal imbalances, BPAs, obesity, secondhand smoke, climate change, pesticides, laptops, tight underwear, hot tubs, and seemingly everyone waiting to have babies until they're 40, countless couples get pregnant every day around the planet. A few months after a positive pregnancy test, couples forget about questions related to fertility, and their focus turns from levels of LH and HCG to dealing with back pain, setting up a nursery, and coming up with a birth plan.

The entire process of getting pregnant, being pregnant, and giving birth makes me think of riding a roller coaster. Usually the first thing you do with a roller coaster is wait in line. Some people end up waiting in a very short line. Maybe they were smart and went on a weekday when schools were still in session. Maybe they have some premium passes that let them cut to the front of the line. Other people have a much longer wait. They go to ride the roller coaster on a holiday weekend in the summer and spend forever in a line snaking around through countless switchbacks.

No matter whether the line was long or short, in the end everyone who gets pregnant ends up buckled into a small car not knowing exactly what to expect. The coaster slowly clicks up the first hill, and you even get a second or two at the top to take in the view and catch your breath. Then the ride hits the tipping point, maybe you scream, maybe you don't, but either way there is no looking back.

9

The Greatest Pain of All

Which type of animal births its offspring through the clitoris?

a. pygmy shrew
b. spotted hyena
c. snow leopard
d. ring-tailed lemur

For my wife, Julie, the pregnancy roller coaster crested the hill and started twisting and turning on an ill-fated road trip. She had recently finished her graduate work in California, and we decided to make the trek from Idaho to attend the commencement ceremonies. A month before the trip, we learned Julie was pregnant. Everything was proceeding smoothly, and we figured 30 hours in the car would not be a problem. In hindsight, it was a bad decision.

If the nausea had kicked in right away, we would have turned around and skipped the graduation proceedings. It did not get bad until the second half of the trip, by which point we were hundreds of miles from home. For reasons that now seem foolish and ill conceived, we decided to take the scenic route back. We drove through the northeastern corner of California instead of setting the cruise control on I-5 and heading home the fastest way possible.

My most vivid memory of the journey is when we pulled into the dusty northern California town of Alturas. We arrived late in the day, and Julie was already feeling green. We found a roadside motel and went to check in. We don't preview motel rooms as a matter of routine, but these were unusual circumstances. Julie's typically keen sense of smell was trending toward bloodhound levels, and the exterior of the motel did not smack of cleanliness. We cracked open the door of our potential accommodations and did a sniff test of the musty room. Julie immediately did a 180 and headed for daylight. She made it safely out and proceeded to puke all over the parking lot. Morning sickness had officially begun.

I politely declined the room, and we ended up in the next place down the road. Motel number two did not work out either, as someone started smoking in the room next to us in the middle of the night. In Julie's heightened state of awareness, I might as well have been puffing cigarettes in bed and blowing the smoke right in her face. After a fitful night, we gave up on resting in Alturas around 4:00 a.m. and hit the road. With many stops and starts and breaks for fresh air, we eventually made our way home and settled into a first-trimester routine.

The nausea and vomiting were only part of the challenge. Figuring out any acceptable food for her pregnant body was also a stumbling block. The most reliable food Julie could keep down early in the pregnancy was oatmeal. Every hour or two, I would make another small portion, and Julie would nibble at it and try not to think about northern California motels. The first trimester diet seems to have imprinted hard on our daughter, Ellie. To this day, oatmeal remains one of her favorite foods.

Sometime later in the summer, I found myself at a car dealership, as we were thinking about trading in our old beater. The young salesman and I got to talking, and it turned out his wife was also pregnant. I mentioned that my wife had been fighting morning sickness and had not gained any weight during the first trimester. I knew Julie's situation was not wildly unusual, but most women put on at least a little weight during the first 12 weeks.

In the vein of "it can always be worse," the salesman's wife had *lost* 20 pounds (9 kg) during the first trimester. She had hyperemesis gravidarum. *Hyper* means excessive, like an excessively busy, *hyper*active kid, or Chewbacca shifting a spaceship into *hyper*drive to go extra fast. *Emesis* means vomiting. An emetic induces vomiting. *Gravidarum* means during pregnancy. Biologists usually reserve the word *gravid* for nonmammalian animals like snakes or spiders when they carry fertilized eggs around in their abdomens. This is one of the rare instances when you see the word applied to humans. So hyperemesis gravidarum is a fancy way of saying someone experienced an excessive amount of vomiting during pregnancy. Again, the car salesman's wife *lost* 20 pounds in the first trimester. I drove home (in the same old beater) thanking my lucky stars we could at least get Julie to keep down oatmeal.

Julie managed to avoid a diagnosis of hyperemesis gravidarum, but the pregnancy did not get much easier in the second trimester. She was still sick around the clock (the "morning" of morning sickness is rather misleading). We pivoted away from bland oatmeal to bland pinto bean burritos as the pregnancy food of choice. Julie did not have any of the classic cravings you hear about accompanying pregnancy. She never sent me off into the night to buy pickles or chocolate or chocolate-covered pickles. She did, however, experience some significant food aversions.

She wholly rejected meat, and garlic became public enemy number one. Her aversion to garlic was so strong, I tried to avoid eating garlicky foods for fear of sending her running to the bathroom. I would sometimes slip up and have a savory burger or spaghetti for lunch. Upon arriving home, Julie's body would sense I had broken the anti-garlic oath. The garlic aversion did not fully abate until Ellie was two or three years old, at which point we were able to return to cooking the way we had prepregnancy, albeit with a lot more mac and cheese in the dinner rotation.

Julie's morning sickness lasted until the start of the third trimester. The discomfort then simply shifted to the standard joys of late pregnancy, including heartburn, back pain, and sleeplessness. Long story short, as it is for a lot of women, the entire process was decidedly not all sunshine and roses. Having a front row seat for all of the suffering made me wonder why the whole trifecta of pregnancy, birth, and nursing is such a complete and total shit-show for humans. It seems like *way* less of a big deal for other animals. Watching a loved one go through it makes you appreciate what an incredible sacrifice it is for a woman. All you can do as a guy is be as sympathetic as possible, make countless bowls of oatmeal, and remember to avoid the garlic fries at all costs.

NVP

I tried, gently, to remind Julie, between bouts of puking, that morning sickness, or nausea and vomiting in pregnancy (NVP), was a sign of a healthy pregnancy. I was aware of this because of a paper published by a friend of mine while we were in graduate school. Sam Flaxman and his advisor, Paul Sherman, compiled evidence from an array of studies supporting the idea of NVP as a mechanism to protect the mother and child during the particularly sensitive period of early pregnancy.

Building on the work of Margie Profet[1] (the author we first met in the menstruation chapter) and others, Sherman and Flaxman presented a compelling case for what they deemed the maternal-and-embryo-protection hypothesis. They argued in their paper in the journal *American Scientist* that NVP "protects the developing embryo from teratogens and also protects both the mother and her embryo from food-borne microorganisms."[2] The first piece of evidence in support of the hypothesis is that the symptoms of NVP and the food aversions often accompanying NVP tend to be most prevalent during the time of the pregnancy when the embryo is most susceptible to developmental errors. The first trimester is a period of rapid cell division when the anatomical and physiological foundations for all the major systems of the body are put in place. It is the most critical period of time during a pregnancy for a woman's body to protect the embryo from potentially damaging chemicals and microorganisms.

One of the common ways the body gets exposed to potentially dangerous pathogens is when bacteria hijack a ride on consumed meat. Meat provides an ideal culture for bacteria. This is why it is best to eat fresh meat and also why it is safest to cook meat before consuming it. Pregnancy resets a woman's tolerance for what her body considers fresh. An aversion to meat is the most common type of aversion experienced by pregnant women, and the aversion is strongest during the first trimester.[3]

Data related to pregnancy outcomes supply the most convincing evidence that NVP is a positive behavior (rather than a sickness). Women who experience NVP are less likely to suffer miscarriages than those who sail through with no NVP symptoms.[4] The maternal-and-embryo-protection hypothesis suggests NVP is a sign that the pregnant body is doing everything it can to keep potentially disruptive chemicals and microorganisms away from the developing embryo or fetus.

Vomiting is a sign of a body on full alert. It may be an overreaction when it occurs during pregnancy, but overreaction goes part and parcel with extreme caution. Let's say, for example, you live out in the sticks and a mountain lion has been roaming your neck of the woods lately, picking off cats and dogs and the occasional small child. You sit on the porch, rifle in hand, keeping an eye out for the cougar, protecting your family. Anytime you hear a twig break or see something out of the corner of your eye you fire off a shot or two, just to be safe. This behavior may not be rational, but it might be how you would respond to a blood-thirsty mountain lion roaming your property. You might kill an innocent rodent or two in the process, but better to smoke a squirrel than to let your guard down and have a puma in the yard.

Adult immune systems are strong enough that we don't have to remain hypervigilant against bacteria at all times. If a few sneak in there on a poorly cooked piece of meat, it is usually possible to beat them back without too much fuss. On rare occasions, the immune system becomes overwhelmed, and the brain and gut take over and throw a meal back up. But for the most part, even when ticking along with only its normal level of activity, the immune system is brilliant at handling rogue invaders.

The embryo, however, does not have a functioning immune system in place yet; the mother is in charge of its protection. Thus, mom has to set the system on full alert. She may be able to survive a bacterial infection, but it is unlikely her developing baby would make it through unscathed. So she resets the security system to an extreme level of surveillance and sits there, trigger-happy, on the porch, in a rocking chair, rifle in hand.

The maternal-and-embryo-protection hypothesis makes one other important prediction about NVP. Its presence should vary globally based on the likelihood of ingesting contaminated food. If there are basically no mountain lions around, then homeowners do not need to be so wary. The likelihood of picking up bacteria via the diet varies considerably depending on where you are on the planet. Some cultures consume considerably more meat, and consequently more contaminated meat, than others. Sherman and Flaxman presented evidence demonstrating NVP is far more common in cultures with meat-heavy diets. There were even a few societies in the studies with zero reported incidence of NVP. People living in those locations were "significantly less likely to have meat as a dietary staple and significantly more likely to have only plants as staple foods."[5]

In addition to meat, the other food products highlighted by the research are those loaded with phytochemicals, including vegetables, coffee, and tea. Upon publication of this type of work, people naturally tend to overreact, and some jumped to the conclusion that meat and vegetables are dangerous for the development of human offspring. Such an overreaction is risky and flawed. If a pregnant woman is able to tolerate them, meat and vegetables are a healthy part of a prenatal diet. These days, modern processing and refrigeration make the likelihood of bacterial contamination low, and many commonly ingested vegetables have undergone generations of artificial selection, minimizing their load of phytochemicals.

Physiological memory is long, however. The evolutionary drive to protect the developing embryo remains strong, even if technology has beaten back the traditional enemies. And although the risk of ingesting contaminated food is at a historical low, it is not as if we have created a world without food-borne illness. People still suffer from food poisoning. Anyone who has lived through a 24-hour gastrointestinal infection knows how miserable the experience can be. It is especially important for a pregnant woman to avoid a bout with a nasty bacterial infection, and NVP is a natural avoidance mechanism. The result is a bland pregnancy diet and months of feeling green on the pregnancy roller coaster. Through it all, however, the extra protection is worth it because a woman who has to fight through NVP typically ends up getting to the end of the ride with a healthy baby as a reward for all her suffering.

ONE MILLION MOTHERS

Let's not get ahead of ourselves here. We're not ready to birth the baby quite yet. There are other, even more serious, dips and turns further along the pregnancy ride. Many of the pregnancy-related problems relate to the previously discussed issue of the human fetus burrowing into the uterus so deeply. It does not dig in aggressively just to snuggle up and get cozy. It digs in all the way next to the maternal blood vessels so it can control the mother's physiology. The fetus subconsciously plays the role of puppeteer, secreting manipulative hormones to try to get the mother to act exclusively in the best interests of the unborn child.

One of the most common problems of pregnancy is gestational diabetes, with upwards of 10% of pregnant women suffering from the condition. All types of diabetes involve disruptions in blood sugar levels, and pregnancy places unique pressure on the mother to provide enough energy for the development of the fetus while still maintaining an adequate supply for her own cells. It is in the best interest of the fetus to divert energy away from the mother because the fetus can use any "stolen" energy to benefit itself. It is counterproductive to imperil the mother's life by stealing too much, but the fetus often diverts enough to make the mother feel miserable and exhausted around the clock.

The fetus accomplishes this manipulation via the placenta and the production of hormones that counteract the effects of insulin. Insulin is the hormone that makes it possible for cells to take up sugar from the blood. Without insulin, sugar just circulates in the blood, making it useless for the cells that depend upon it as an energy source. With insulin rendered less effective during pregnancy, more of the mother's sugar remains in the blood, which means the fetus has access to more sugar. The placenta is not fully formed until around the 12th week of pregnancy. This explains why the early embryo is not able to manipulate the blood sugar levels of the mother to the same degree it can later in the pregnancy.

Like NVP, gestational diabetes shows some fascinating geographical trends. One great way to determine where on the nature–nurture spectrum a condition like gestational diabetes falls is to work with a melting pot of a population. The best place for such a study is somewhere where people live in similar ways, but where they are all descended from different, diverse populations. Places like London, Paris, Hong Kong, Toronto, and New York City meet that description very well.

A group of researchers dug into NYC pregnancy and birth data and compiled results from more than one million women over a nine-year period.[6] We've seen some large sample sizes in studies up to this point, but gathering data from one million women takes the prize. The researchers of this one-million-woman study discovered ethnicity plays a significant role in dictating the likelihood that a pregnant woman will develop gestational diabetes.

It all comes down to how many carbs your ancestors ate. Ancestral Europeans, starting around 10,000 years ago, shifted from a hunter-gatherer lifestyle to grain-based agriculture and a diet filled with carb-heavy grains like wheat. Shortly thereafter, around 8,000 years ago, they started to consume sugar-rich milk from domesticated animals. Ancestral women in South Central Asia, in contrast, ate foods with more moderate levels of carbohydrate, like fish and unprocessed rice.[7]

The results show that the risk of developing gestational diabetes is *inversely* related to historical access to carbs. Only 3.6% of women of European descent in the study developed gestational diabetes. By comparison, the condition occurred in 14.3% of women of South Central Asian descent.

It is easiest to understand these trends by pulling on some Darwinian pants and analyzing the numbers in the context of natural selection. In the modern world, a woman with gestational diabetes is more likely to need a C-section. The diversion of sugar and energy to the fetus leads to a bigger baby, which is, naturally, more difficult to birth. A woman with gestational diabetes is also more likely to suffer from other pregnancy- and birth-related complications, including preeclampsia (discussed in the next section), and she has an increased risk of tearing and hemorrhaging at birth.[8]

European women are protected by the history of their population. Roll the clock back a few thousand years and any European women prone to gestational diabetes would have been in trouble. Without access to modern medical care, on their carb-heavy diets, many of them would have died during pregnancy or childbirth of complications related to gestational diabetes. Genes predisposing women to the condition were weeded out of the population, which is why those genes are seen less frequently in modern women of European descent.

The conditions were not the same, historically, for a woman in South Central Asia. A woman there was not eating as many carbs, leading to selection *for* genes that increased blood sugar during pregnancy because it would have shuttled the necessary energy to the developing fetus. Because her ancestors had a different back-story than Europeans, a modern woman of South Central Asian descent is likely to possess genes predisposing her to a gestational increase in blood sugar.

Roll the clock forward a few thousand years and now everyone in NYC has access to more than enough sugar. There are buckets and buckets of sugar available with every snack or meal. It is no way to treat a pregnant body, but given most people's diets these days, in all likelihood a pregnant woman living in NYC is sucking down either a soda or a sugary coffee drink every day, many of which contain more than the *daily* recommended amount of sugar.

For all women, consuming excessive amounts of sugar during pregnancy is potentially dangerous. A pregnant European-American woman is more likely to get away with such behavior, however, because of the strong selection against gestational diabetes in her past. A pregnant Bangladeshi-American woman gets burned by it and develops gestational diabetes because of the strong selection for genes that increase blood sugar in her past. An astonishing 21.2% of women of Bangladeshi descent in the study developed gestational diabetes. Even if Bangladeshi-American women drink only water and kale smoothies during pregnancy, they may still develop the condition because of their unlucky evolutionary history. Once again, you cannot outrun your past.

THE DELICATE BALANCE

Developing human fetuses clearly need *a lot* of energy. The primary reason for this is to feed and grow the expansive human brain. By the third trimester, when the brain is growing dramatically with each passing day, some 60% of the energy the baby gets from the mother is diverted to the brain.[9, 10] Sixty percent is an absurdly high number, with most other mammals dedicating only around 20% of their fetal nutrition to brain development.

The developing fetus needs more than just sugar; it also needs loads of oxygen. Without oxygen the body cannot unlock the full energetic potential of sugar. It is why our bodies run out of gas when we push them into oxygen deficit (like when we vigorously exercise) and why they really run into trouble if they are deprived of oxygen (like when someone drowns).

Human fetuses need a uniquely large supply of sugar and oxygen because they are trying to grow a massive brain. The maternal blood carries both of those reactants (sugar and oxygen), and fetuses have intrusive tactics to make sure they get what they need. First, as has already been discussed, the fetal portion of the placenta aggressively invades the maternal tissues. It digs in much deeper than it does in other mammals. An unborn puppy can derive the nutrition it needs from its mother without rooting in to such a degree that it practically kills mom. The only way human fetuses can get enough sugar and oxygen is to push the envelope with highly invasive placentation.

In a final act of aggressive, selfish manipulation, the fetus often causes a spike in the mother's blood pressure. By increasing the mother's blood pressure, the fetus is able to enhance the number of nutrients it can leech from the mother. A small increase is common and not particularly dangerous. The mother, after all, has a significantly increased blood volume (it can rise during pregnancy by as much as 50%), and there needs to be enough pressure to support growing an entirely new organism inside the womb. There is a delicate balance, however, between a slight increase in blood pressure and a dangerous elevation of blood pressure. It is not uncommon for the balance to be thrown off. If the scales tip too far, the mother's blood pressure rises to an extreme level and creates the condition of preeclampsia.

The situation is unique to humans. Exactly what triggers it (besides, you know, growing a new life inside you) is still a mystery. Preeclampsia occurs in 10% of pregnant women and in the most serious cases progresses to a state where organs start to shut down, leaving the lives of the mother and the baby hanging in the balance. It is a significant cause of perinatal mortality across cultures and ethnicities[11] and is the leading cause of maternal morbidity in developed countries.[12]

THE PAIN FOR EVERYONE

In addition to potentially fatal conditions like gestational diabetes and preeclampsia, there are other more common pains of pregnancy. We are bipedal animals, and pregnant women carry babies around in bodies largely made for a life on all fours. The body has been beautifully jury-rigged to work on two feet, but an event as dramatic as pregnancy shines a bright spotlight on the flaws of the Johnny-come-lately (or in this case, Jenny-come-lately) bipedal life. As previously discussed in the chapter on back pain, a woman's center of gravity shifts forward during pregnancy. A bipedal pregnant body effectively has two options: fall over or suffer from back pain. The back takes on the responsibility of countering the change in the center of gravity. The pregnant spine takes on even more than its already high degree of curvature, and significant shearing and pain naturally result. The back pain gets worse and worse deeper into pregnancy because the little bowling ball keeps growing and growing and growing.

All the fetal growth may add to another commonly cited complaint of pregnancy—heartburn. The increasing size of the human fetus throughout pregnancy places increased pressure on the stomach and, in particular, the lower esophageal sphincter (or LES), which sits between the stomach and the esophagus. The job of the LES is to keep the acidic contents of the stomach inside the stomach and out of the esophagus. Heartburn occurs when some stomach acid slips past the loosened sphincter and splashes up into the esophagus.

The etiology of pregnancy heartburn remains somewhat of a mystery, but there seems to be a consensus that increasing levels of estrogen and progesterone play a significant role. The levels of both those hormones spike tremendously in pregnancy. Estrogen concentrations elevate during pregnancy to 30 times the typical amount circulating in a nonpregnant woman's body. Thirty times! It is incredible that women can tolerate such a dramatic shift in their physiology and continue to go about their daily lives. We would need a lot more room in prisons if the analogous hormonal change happened in men and their testosterone levels increased 30-fold over the course of several months.

Another random piece of evidence exists in support of the connection between gestational heartburn and increased hormone levels. In addition to their other varied roles, steroid hormones, including estrogen, modulate hair growth. A long-held wives' tale about pregnancy suggests women who suffer from heartburn are likely to have newborns with thick heads of hair. Unlike a lot of pregnancy wives' tales (like the laundry list of ridiculous ones about predicting gender, all of which are correct 50% of the time), it turns out this one may hold water. Researchers tracked women through pregnancy and monitored their degree of heartburn and how hirsute their babies were at birth.[13] Lo and behold, most women in the study with moderate to severe heartburn birthed hairy little kids and most women with no heartburn birthed hairless wonders.

So the LES is already in a loose, accident-prone state because of the high hormone levels of pregnancy, and then you go and put a watermelon in the abdomen. It is clearly a recipe for trouble, and the majority of women experience some heartburn in addition to the litany of other pregnancy symptoms. There is light at the end of the heartburn tunnel, however. For most women, the heartburn cuts out as soon as the kiddo comes kicking and screaming into the world.

THE MICHAEL JORDAN OF PAIN

All the pregnancy pain and misery are only a preamble to the biggest nightmare associated with human reproduction: giving birth. People typically evoke the pain of childbirth to compare levels of suffering when someone has a different painful experience, like a compound fracture, kidney stones, or a root canal. Childbirth is the gold standard of pain. It is the Mount Everest of pain. It is the Michael Jordan of pain. It sets the bar for all other types of pain regularly experienced by humans.*

It is not like this in other animals. Parturition (a fancy way of saying childbirth) is much less of a production in just about all other mammals. Most, including other great apes, labor for a couple of hours, push the baby out, give it a few licks, and that's that. Humans labor *a lot* longer. A study of more than 1,600 women from Nordic countries kept track of labor lengths and demonstrated averages of

---

*There are obviously rare events capable of generating a truly unimaginable level of pain that would exceed even childbirth. Getting mauled by a grizzly bear, losing a limb in a freak train accident, things of that nature. Of the normal events, however, that happen to a good percentage of people in the typical course of a lifetime, it is hard to think of anything more painful than childbirth.

14 hours for first-time moms and 7.25 hours for mothers giving birth for at least the second time.[14]

Not only is the birth process long and arduous in humans, it is downright dangerous. Before about 1935, the odds of dying while giving birth were around 1%. Most women had more than one child. Two kids meant a lifetime 2% chance of dying during birth. Five kids meant a 5% chance of dying. One percent adds up in a hurry. Historically, an incredible number of women died during childbirth.

The two most significant risks during childbirth are severe bleeding and infection. Remember how I mentioned earlier that a woman's blood volume can rise as much as 50% during pregnancy? One reason for that is to support the growing fetus. The other reason is so that the body can survive the vast amount of blood lost during birth. The average amount of blood lost during a vaginal birth is 500 milliliters, and a C-section usually leads to the loss of more like a liter of blood. In either case, it's a veritable pool of blood lost during birth.

Even if a woman manages to survive the incredible loss of blood, she is still not out of the woods. The risk of the mother dying during parturition took its first significant dip in the 20th century with the advent of antibiotics. Doctors could finally control the types of infections that were commonplace after the uniquely wounding human birth process. Nearly 100 years after the discovery of antibiotics, many portions of the globe still do not have ready access to modern medicine. In many parts of the planet women still die during childbirth at rates approaching historical numbers.

Another issue around the topic of pregnancy-related deaths is that, for reasons not fully understood, the problem affects certain ethnicities more than others. According to data from the CDC, Black and Native American women are two to three times more likely to die during childbirth than white women.[15]

Birth is unquestionably brutal for humans. Even with antibiotics and much more advanced obstetric care, hundreds of thousands of women still die each year from complications associated with childbirth. For women who survive, the event often permanently changes their pelvic anatomy. The symptoms are categorized under the umbrella term *pelvic floor disorders* and, as written about in an article in the journal *Current Opinion in Obstetrics & Gynecology*, include "stress urinary incontinence (SUI), overactive bladder, pelvic organ prolapse and fecal incontinence."[16] Nearly one out of every four women in the United States suffers from some type of pelvic floor disorder, and some reports place the global prevalence as high as 46%.[17, 18*] It is an astounding number to consider and makes the topic feel like one of the most ignored issues in modern medicine. These are the topics no one has the guts to talk about with expectant mothers.

How did we ever get to this point? Why do women have to literally risk their lives to bring a new life into being? Why is human birth so ridiculously difficult and problematic?

---

*The study included parous and nulliparous women (women who have and have not given birth, respectively). Of the 1,961 study participants, 23.7% suffered from at least one type of pelvic floor disorder. Not surprisingly, the condition was least common in nulliparous women (12.8%) and increased in frequency with the number of children birthed (18.4%, 24.6%, and 32.4% for women with one, two, and three or more deliveries).

SWAB THAT BABY

It all starts with our heads. Our broad shoulders are also a problem, but not one that seems to have gotten as much attention as our large heads. There are many hypotheses and debates about the origins of the difficulty of human birth. Most of it boils down to one entirely unassailable fact: humans have giant brains encased in large skulls. It is no picnic birthing those huge noggins. It was the biggest hang-up for getting our daughter, Ellie, out into the world.

Julie's labor started on a Sunday night at 10:00 p.m. At the time, we were watching TV and Julie moved from the couch to an exercise ball to manage some of the early labor discomfort. Her contractions were roughly 10 minutes apart, and we figured we'd have a few hours at home before heading to the hospital.

As anyone who has considered a birth plan knows, there are countless different methods, strategies, and techniques for giving birth. At one extreme, there are women who give birth in their homes in warm tubs of water with doulas holding their hands while songs stream from gentle birthing playlists. At the other end of the spectrum are women who have their C-sections scheduled months in advance. They receive epidurals upon arrival at the hospital, and everything runs on a schedule. We aimed for somewhere in the middle of those two extremes. We chose to have Julie give birth in a hospital, but instead of an obstetrician (OB) we used a midwife. An OB would be on call if Julie needed surgery, but the plan was for Julie and the midwife to run the show. My role was to fulfill the important job of pacing back and forth.

We chose our particular midwife based largely on the desire to avoid a C-section. C-section rates have skyrocketed in the United States, with 31.9% of kids now born via Cesarean delivery.* At the time, our midwife had delivered 250 babies, and only four of those births had ended in a C-section. We liked the sound of those odds much better than nearly one out of three.

Evidence has piled up in recent years about the immunological benefits of a vaginal birth. If kids come out the natural way, they receive a nice coating and inoculation of mom's bacterial flora as they pass through the tight birth canal. Their immune systems develop differently (especially in the early years), and multiple studies have shown that children born vaginally are less likely to develop asthma, allergies, and some autoimmune disorders like celiac disease and type 1 diabetes.[19]

An OB may even go so far as to swab a newborn from a C-section patient with the goop from the mother's vagina in an attempt to simulate the process of vaginal birth.[20] And yes, *goop* is the formal medical term they use in those settings. Getting slimed in this way is not as good as the real deal, but it is better than not getting anything at all. If a baby doesn't pick up a stock of natural bacteria on the way out, the first bacteria their body gets to know are those lying around the hospital, and that can be a tough start for the education of the immune system.

---

*The CDC has compiled the data. You can find it by googling CDC method of delivery.

TWO HOURS VERSUS TWO DAYS

I cannot recall all the details from our prenatal visits, but I do remember asking one specific question. I wanted to know, once labor started, when we should head to the hospital. Our midwife told us it was time to hop in the car if one of two events occurred: if Julie's water broke or if she was experiencing sustained contractions less than five minutes apart. At the time, those seemed like straightforward instructions we would have no trouble following.

We were binging *Game of Thrones* when Julie experienced her first contraction. We figured we'd watch a couple more episodes and then it would be time to go to the hospital. We probably should have chosen something a little more Zen. Brutality and fire-breathing dragons may not have been the best images to set the tone for giving birth.

As it turns out, we could have watched multiple seasons of the show at home. Nineteen hours later Julie's water had not broken, and her contractions were still . . . 10 minutes apart. They were becoming more painful, which was the only sign of progress we seemed to be making. We called our midwife, and she had us meet her at the hospital so she could check on Julie's status. There we learned Julie's cervix was dilated to all of three centimeters. Three centimeters would have worked out fine for birthing a chimpanzee, but Julie was not trying to birth a chimpanzee. Chimpanzee brains are only about one-third the size of human brains, and a mother chimp's cervix needs to dilate to just over three centimeters. As anyone who has given birth to a human baby knows, the human cervix needs to get to 10 centimeters before it is time to push.

Another night passed, and by the morning Julie was up to six centimeters. By noon she was finally up to 10. A few more hours of pushing and our precious bundle of joy was almost ready to come out. Her head was visible, and with each round of pushing it would come close to clearing the opening before being sucked back inside until the next round of contractions. About when a C-section appeared to be inevitable, her head finally passed through the vaginal opening. After all the work to clear her head, the rest of her slipped out like it was nothing. I've never felt more relieved in my entire life.

Ellie was not a large baby. At birth she was 6.6 pounds (3 kg) and 19 inches (48 cm) long, both of which are a bit below average but well within the perfectly healthy range. She did, however, have quite a large head. She came out without a C-section only because our midwife stayed with us for 24 hours and because Julie is one very stubborn warrior. Most other primates labor for two hours; Julie labored for nearly two days.

ONE TIGHT CAVE

Trying to understand why human birth is so difficult makes me think about caving. I am not an experienced caver by any means, but I have been in caves a few times and know it can get awfully tight when slithering through underground tunnels. Fundamentally, only two factors dictate how tight and claustrophobic conditions will become while caving. The first is the shape of the spelunker. Obviously, a smaller individual will have an easier time navigating through narrow spaces. When viewed in this way, a human baby is simply too big for the cave.

The second factor is the shape of the cave. It does not matter how big the caver is as long as the cave has plenty of room. The human birth canal is one exceedingly tight cave. Mothers who give birth to babies with large heads or broad shoulders, or God forbid both, are in for one hell of an uncomfortable experience getting the kid out to daylight.

The shape of the pelvis dictates the shape of the cave. In addition to providing the passageway for birth, the pelvis also serves as the site of attachment and anchor point for bones and muscles that promote locomotion. When human ancestors became bipedal, the shape of the pelvis gradually changed in ways promoting upright locomotion on two feet. Prior to this major change, the muscular abdominal wall supported all the weight of the viscera (or guts). But up on two feet, the pelvis took on the job of supporting the abdominal organs. As a consequence, it became more bowl shaped over time. The bowl shape changed the nature of the pelvic inlet (the start of the birth canal) and made for a tighter birth canal compared with quadrupeds.

Even without getting a view of the twists and turns of the inner passageway, just by looking at a skeleton it is possible to see how giving birth has placed significant selective pressure on the shape of the female body. The angle between the two hip bones, referred to as the subpubic angle, is significantly wider in females than it is in males. The difference is obvious if you have a skeleton of each gender at your disposal.

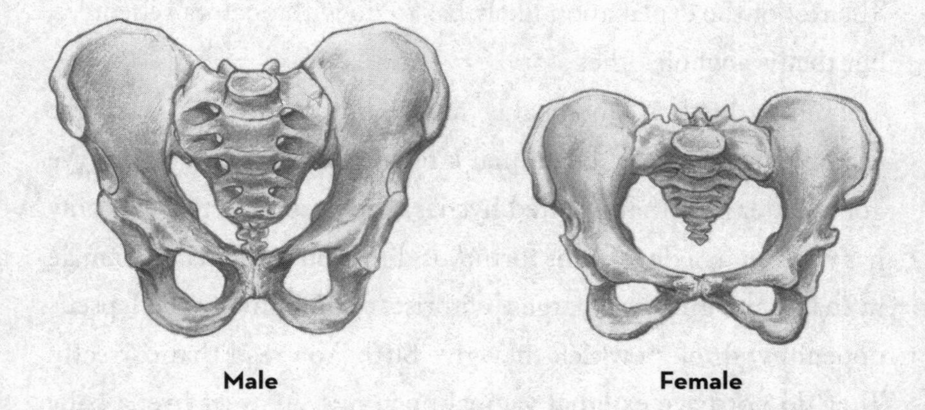

Evidence suggests birth in human ancestors has been difficult for at least several million years. A research group led by Jeremy DeSilva (the paleoanthropologist we first met in the foot chapter) reconstructed the birth process using fossil evidence in *Australopithecus afarensis* (Lucy's species) and found the fit was probably tight in the newly bipedal species, with their bowl-shaped pelvises and broad shoulders, even before the brain swelled up to its current size.[21]

The later dramatic increase in the size of the brain would have then compounded the issues. Modern lifestyles have added the cherry on top of the birth-torture sundae. Many pregnant women now have access to greater numbers of calories in their diets than at any point in human history. Consequently, the selfish fetuses are able to divert more calories to themselves than ever before. Calories turn into pounds, and modern women get the joy and bragging rights of birthing the largest babies in the history of humanity. Babies sometimes grow so big they cannot come out vaginally, which is at least part of the explanation behind the increase in C-section rates. The rest of the explanation likely has to do with doctors' schedules, but that is a whole other story.

Human birth is undeniably rough, but there is at least one example out there in the animal kingdom that is even worse. We got a better deal than spotted hyenas. Their reproductive anatomy has evolved in a downright torturous direction for females. Female hyenas have greatly enlarged clitorises (sometimes called pseudopenises) through which they give birth. You read that correctly. They do not have external vaginal openings. A sweet hyena baby comes out through an opening in the mother's *clitoris*. A good deal of tearing is involved, and it is not uncommon for first-time mothers to die in the process. So, yeah, it could always be worse.

BIRTH PARTY

Misery loves company, as the saying goes, and birth in humans is so miserable that having company has become a normal part of the operation. Thinking back to Ellie's birth, in addition to Julie and me, multiple nurses and our midwife were in the room. This socialization of human birth is not unique to hospital births. Even in cultures in which birth takes place far away from sterilized hospital rooms, expectant mothers universally receive assistance from family members and friends.

Human babies are usually born in an orientation that makes it difficult for the mother to cradle the child. The obvious solution to the problem is to have close friends or family nearby.

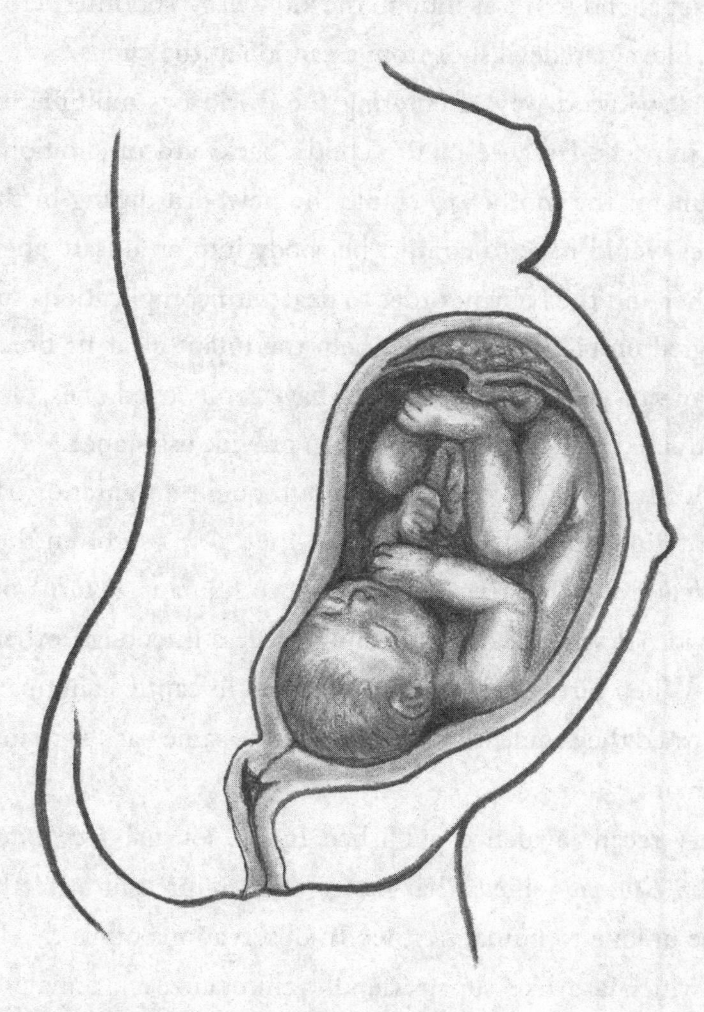

Birth has become a group event for good reason. Human babies make a twist when passing through the birth canal. As a result, babies usually have an anterior occiput presentation, meaning they come out headfirst and are facing toward the mother's back upon expulsion.[22] Expulsion is a funny way to put it, but it is the word often used in the literature to describe *the* moment. Expulsion makes it sound to me as though the kid will go shooting across the room, like a daredevil shot from a cannon at the circus.

This awkward way of entering the world has multiple immediate impacts. Because of the child's backward orientation, it is difficult for the mother to cradle the newborn during birth. The mother would have to contort her body into an unsafe angle for both her and the baby in order to deal with complications such as a tangled umbilical cord or to help the infant clear its breathing passageways. It is helpful, then, to have some loved ones, or a few well-trained professionals, around to provide assistance.

Until relatively recently, scientists thought an anterior occiput presentation was a uniquely human trait. As is so often the case with features we believe to be special to humans, it turns out we just had not witnessed the birthing process in enough other animals. When birth was carefully observed in captive chimpanzees, we learned they come into the world in the same backward manner as humans.[23]

Very recent evidence published in the journal *Evolution and Human Behavior* shows that having helpers present at birth may not be unique to humans either. In observations of captive bonobos (which are like super friendly chimpanzees), females were seen attending births and even helping out during the "expulsive phase."[24] Although the attendance of helpers in bonobos is not obligatory (other observations have made it clear bonobos are fully capable of giving birth on their own), bonobos are gregarious like humans and, like humans, apparently also appreciate company during a trying event.

The similarities between bonobo and human births diverge when it comes to attitudes toward the placenta. We asked the hospital to save the placenta following Ellie's birth. We brought it home and buried it beneath a dogwood tree we planted to mark the occasion.

I can see where such a celebration of new life may be too touchy-feely for some, or possibly too bloody, but it's not like we sat around and ate the placenta together. Oh yes, most mammals celebrate giving birth by eating the afterbirth.[25] A mother bonobo from the aforementioned study shared it with her helpers, which I thought was a nice touch. Placentophagy is gaining traction in some human circles, and the debate over the pros and cons of placenta consumption is very much alive on the internet. I'm going to steer clear as I really don't need a recipe for placenta burgers or placenta lasagna. I am sure they are both best served with a glass of very fresh, warm milk.

LIMP AND HELPLESS

If you could get other primates to comment on human birth, I think they would make a few observations about how the process is distinctly different from their own. For starters, it would seem ridiculous to any lemur, monkey, or ape that it takes a human the better part of a day to push a baby out. I'm sure they would also be surprised by how much screaming and trauma is involved. The explanation behind these first two issues does not need to be overthought. A pregnant woman attempts to pass a large object through a very small tunnel, and that is not an easy or painless thing to do. Human birth makes for the perfect storm in which our two most unique features, bipedalism and huge brains, work together to make the experience painful and challenging.

Other primates would also probably take note of how completely and totally helpless our infants are. The explanation for our uselessness as newborns has to do with the size of the infant brain compared with the size of the eventual adult brain. A newborn human baby's brain is only 30% of the adult size. That is not to say our newborns' brains are small. In fact, they are the largest of any of the great apes. They are *relatively* small, compared with how big they will eventually become. Other primates grow their babies in the womb until their brains are at least 40% of their final size. Some get their brains up to 50% of adult size. The extra 10 to 20% makes a significant difference. It is the difference between a primate who comes out basically raring to go and another who comes out limp and helpless.

Altricial infant animals are those with limited to no skills; infants who are more mobile and mature are called precocial. Those two terms are not discrete options but rather two ends of a spectrum. Far on the altricial end of the stick are animals like rats and wolves. They emerge deaf and blind and are totally helpless at birth. On the other, precocial, end are animals like horses and giraffes, which are trotting around shortly after birth.

Human infants are difficult to place on the spectrum. On the one hand, we have large brains that would suggest a precocial nature. On the other hand, those brains are *relatively* small compared with their eventual adult size, making us rather limp, if you will, because of our limited development at the time of birth.[26] Instead of getting hung up on where we are on the altricial–precocial spectrum, the more pertinent question is, why are human infants born with brains that still require so much growth?

The answer is not simply that we give birth early compared with other animals. Our gestation is actually long for an animal of our size, but we do give birth at a point when the brain is less developed relative to that of other animals. Why? Why not leave the bun in the oven and let the brain reach a size where a newborn could be more independent at birth? It just really feels like the bread needs to stay in for a bit longer. It's like we pull it out while it is still all gooey and uncooked in the middle.

The obvious and short answer is that when you are growing a sizable brain, it is going to have to come out sooner than is ideal. But again, why? What is the limiting factor?

HAM AND EGG

The obstetric dilemma (OD) is the explanatory hypothesis that dominated the discussion until recently. The OD suggests there is a trade-off or compromise between encephalization (brain size) and bipedal locomotion. Wide hips facilitate birth, but hips can be only so wide before bipedal locomotion is impeded, or so the logic goes. Per the OD, a newborn is born underdeveloped because if it grew any larger inside the mother it would not be able to come out.

The idea is easy to visualize and understand, and for those reasons, it became the accepted explanation for both the difficulty of human birth and the helpless nature of human infants. What was missing was convincing evidence to support the hypothesis. Sometimes an idea is so clean and straightforward it is allowed to survive for a while without much evidence. It seems like the earth *should* be the center of the universe, so everyone thinks it is, until someone collects evidence contradicting the original idea. Human birth is uniquely painful, humans are uniquely bipedal, the two are probably related, or so the thinking went.

A biological anthropologist at the University of Rhode Island named Holly Dunsworth has been going all Galileo on the OD in recent years. She collaborated with researchers at Harvard University and arrived at a different idea. Their data show, from a biomechanical perspective, how wider hips do *not* decrease the efficiency of movement.[27] In other words, a broader pelvis does not make women's strides any less effective than men's. In fact, she argues, there is enough natural variability in pelvic dimensions that if the shape of the pelvis were the limiting factor in dictating fetal size, humans could gestate their offspring until their brains were 40% of the eventual adult size, similar to chimpanzees. It would require only a three-centimeter increase (just over an inch) in the size of the pelvic inlet, which is well within the natural skew already seen in women.

Instead of pelvic dimensions as the limiting factor, Dunsworth has a different explanation for the timing of human birth. She argues the limiting factor is maternal metabolism. The maximum metabolic rate considered sustainable in humans is somewhere between 2.0 and 2.5 times the basal metabolic rate (BMR). Pregnancy pushes the metabolic envelope like no other sustained human activity. Six months in and a pregnant woman is already up to nearly 2.0 times her BMR, with three months left to go.

The crux of Dunsworth's hypothesis is that the baby has to come out once the mother can no longer support it metabolically. Once it reaches the point where its demands outstrip the mother's ability to provide for it via pregnancy, labor begins. In mammals with relatively smaller brains (meaning all other mammals), that does not happen until the fetus has grown to a point where its brain is 40% of adult brain size. Even then, because their brains are smaller, other mammals have plenty of room to get out during birth. In humans, because the growing brain demands so much energy, mothers reach the energetic limit when the fetal brain is only 30% of adult size. By that point, because the fetal brain *is* large (even if it is small relative to its final size), human mothers are also pushing the pelvic limit.

Dunsworth and her group have named the idea the energetics-of-gestation-and-growth hypothesis, or EGG. She wanted to call it the HAM (humans are mammals) and EGG hypothesis but thought better of it. There is some resistance to this new hypothesis. She has not been convicted of heresy like Galileo, but there are some who are still tightly wed to the original ideas of the OD hypothesis. They make the argument that if early birth has to do with metabolism and energetics, why is the fit between the baby and the birth canal so frustratingly tight?

Dunsworth presents several clever counterarguments to this challenge. First, she notes that although width is obviously an evolved trait (as evidenced by the gender differences in pelvic shape), difference in pelvic shape does not automatically define it as the trait that dictates when a baby must be born. After all, most other primates give birth well before they reach their pelvic limits, meaning there are likely different factors at play in other species as well. Dunsworth would argue that those other factors are metabolism and energetics.

Second, even though the fit is undeniably tight and human birth is unquestionably laborious, it still works well enough. There are more than seven billion of us running around, after all. It may be a pain, but the difficulty does not typically prevent women from giving birth.

Lastly, just because two anatomical features make for a tight fit does not automatically mean one is the cause of the other. Dunsworth notes how a finger also fits perfectly into a nostril, but no one would argue for that being anything more than a coincidence.*

---

*In addition to several journal articles about the EGG hypothesis, Dunsworth has written about the topic multiple times for a lay audience on the evolutionary blog *The Mermaid's Tale*.

The conversation is far from over. The OD supporters may volley back and discover other features of the skeleton or novel aspects of locomotion that force a compromise between bipedalism and birth. Thus far, the EGG supporters have made arguments related to locomotory efficiency and economy, but it is possible that broader studies looking at other features of movement like speed and balance will shed new light on the conversation. Stay tuned.

NIPPLE CREAM

After the exhaustion of pregnancy and the trauma of birth, there is finally a new baby. Hooray! Now let's get the larval thing attached to your nipple! Ready, here we go . . . Ahh! Holy shit, is it supposed to hurt this much?!

Of all the pregnancy, birth, and newborn experiences, breastfeeding caught me most off guard, and I wasn't even the one doing the nursing. I'd seen how miserable pregnancy could be for some women, and the difficulty of birth is not exactly a well-kept secret. Granted, I was not expecting 43 hours of difficulty. The pain of nursing, however, was a surprise to me. In my mind, Ellie was going to latch on and peacefully drink while staring into mom's loving, brown eyes with her own loving, brown eyes. When she'd had her fill, she would let out a sigh, maybe a cute little burp or two, and then she would drift off to sleep.

As it turns out, there is often more sputtering, chapping, and crying than contented staring, cooing, and sleeping. Every new parent must have a moment when it hits them how much their life has changed. For me, the moment came while drearily wandering the aisles of a grocery store in the dead of night looking for nipple cream. That's not the sort of thing you do before kids, or, at the very least, it would be awfully weird if you did.

Most women experience pain during breastfeeding. In some cases, the pain becomes intense. Typically, the pain is most severe during the first week,[28] and many women stop breastfeeding when the pain becomes overwhelming. The discomfort puts mothers in a difficult spot because the benefits of breastfeeding are well established at this point. Children who breastfeed get a big immunological boost by receiving antibodies directly from their mothers via breastmilk. Breastfed children are less prone to colds and ear infections and have healthier GI tracts. They have a decreased risk of developing obesity and are less prone to certain autoimmune disorders such as type 1 diabetes and multiple sclerosis. Mothers also benefit from breastfeeding, with evidence showing that mothers who breastfeed decrease the odds they will one day develop breast or ovarian cancer.[29]*

The American Academy of Pediatrics recommends children be exclusively breastfed (meaning zero other food) for the first six months of life and then breastfed with the introduction of other

---

*The NIH has compiled information, references, and links about the benefits of breastfeeding, which can be found by googling benefits of breastfeeding NIH.

foods until they are at least one year old. With the benefits no longer a secret, most mothers in the United States (more than 80%) start off breastfeeding, but only 35% are still nursing their babies by the end of the first year.*

Researchers have tried to figure out why so many women stop breastfeeding after a short amount of time. One of the most common causes of early weaning is nipple pain. There is no shortage of work on the subject. Journal articles about nipple pain currently litter the surface of my desk. A common theme in the nipple-pain literature is a focus on the positioning of the child. This is where the nipple and nursing story ties back to the birth story and the helpless human infant. Because they are so motor challenged at birth, human infants are not able to get themselves into the proper position for nursing without help.

Once again, mom has to do all the work. Nursing is so different for animals where the young are born more precocial. I know because I have watched an uncomfortable number of videos of nonhuman animals breastfeeding on YouTube. The most telling are the ones with pigs. One-day-old piglets run around and fight over nipples while the mother lies there. You would hardly even know she is alive as the newborn piglets squeal and stumble over each other, trying to get in the best position for nursing. The piglets do the positioning work, and by all accounts, they are quite effective at it. Don't get me wrong, I'm betting nursing is still not entirely pleasant for the mother pig, but it does seem as though

---

*The numbers can be found by googling CDC breastfeeding report card.

the piglets carry the bulk of the burden of getting the operation to work.

Scale up evolutionarily to a chimpanzee and the young are also able to get themselves where they need to be for breastfeeding. Having a 40% brain instead of a 30% brain makes all the difference. A baby chimp may not be running around like a piglet, but it is able to hold on to its mother entirely by itself. A mother chimp can even walk around with her baby holding on. In one of the videos I watched, while the baby nursed, the mom casually checked out her fingers, gave her knee a scratch, nuzzled the kiddo, and seemed totally chill about the whole thing. No sputtering and awkwardness, and if her nipples were hurting, boy, the mama chimp did not show it.

Getting a human baby into the proper position for breastfeeding is not nearly as simple. The American Pregnancy Association has no fewer than 13 bullet points on their website about how to get a good latch established. Do this, don't do this, align this, aim this, hold this, tilt this, wiggle this, tickle this, grasp here, not there, and on and on and on. With midwives, lactation consultants, nurses specializing in lactation, and organizations like La Leche League International, an entire lactation community exists to help mothers navigate the nursing waters. Getting off on the right foot is key, because once the chafing and chapping set in, the nipples may not be able to recover in time to make nursing feasible.

ONE MORE WRINKLE

I think most of the difficulty of nursing boils down to the limited neuromuscular development of newborns. They reach a point where their immense, energy-demanding brains force birth to occur, but when they come out, they can't even hold their stinkin' heads up, let alone get a solid grip on anything. Their loose and limp bodies make nursing difficult and painful. It is important for women to know about the pain of nursing up front, because the unexpected nature of the pain is one of the factors that eventually drive many women away from breastfeeding. If, beforehand, you know something is going to be tough, and potentially even miserable, it gives you time to prepare mentally. There are many support systems available to new mothers *after* birth, but more candid and open communication about nursing *before* birth would help some new moms be able to fight through those first difficult weeks.

It's not all pain and agony. There are little snapshots of joy. You get that perfect baby photo. The baby finally sleeps for more than an hour. You slowly get it all figured out. The burping, diapering, swaddling, and everything else start to click. Eventually, for most women, intense pain during nursing subsides. There is often still pain, but dull pain beats intense pain every time. There is finally light at the end of the long tunnel of discomfort that is pregnancy, birth, and nursing. But of course, in that moment, when it all feels like it is coming together, the baby always has a new wrinkle up their sleeve. Just when nursing starts to feel like second nature the baby has an incredibly fussy few days, and you look down and see the new anatomical feature they've been working on . . . teeth.

Conclusion

A Whale in the Water

Among mammals, humans have the largest brains relative to the size of their bodies. Which animal comes in second place?

a. dog
b. chimpanzee
c. elephant
d. dolphin

The cutting of baby teeth brings us full circle in the story of the evolutionary origins of the aches and pains of the human body. In short order the primary teeth become loose and give way to secondary teeth. Many children will have their wisdom teeth pulled and need braces to fix the disconnect between the human jaw and teeth. Some kids will also have blurry vision and their parents will get to add regular optometrist appointments on top of visits to the orthodontist.

A lifetime of anatomical land mines awaits, and many of the pitfalls, like the dangerous intersection of the trachea and esophagus, have their roots in our ancient past. As kids run around on two feet in our uniquely human way, the lucky ones end up with nothing but the occasional scrape or bruise. Unlucky kids have an arch that doesn't develop correctly or an injury-prone knee and spend their formative years in and out of operating rooms. They hobble to school in unwieldy boots and leg braces as surgeons try to convince their limbs to work bipedally despite the pages and pages of quadrupedal blueprints built into human DNA.

For females, menstruation is then right around the corner with all its attendant aches and pains. As we move into our 20s, 30s, and beyond, a fresh set of problems awaits. Nearly everyone eventually takes to wearing glasses or contacts, and some people start snoring. Others will suffer from more serious issues like sleep apnea or a failing lower back. Women carry a heavier load of the evolutionary anatomical burden with the unique pain they endure during pregnancy, childbirth, and nursing.

We started in the introduction with a question: why is the human body so uniquely prone to aches and pains? In some instances, the answers were logical and more obvious, like in the case of back pain. You cannot take a horizontal spine and make it vertical and not expect a few problems. In other cases, the answers were more complex and confusing. For example, we still do not have a complete picture of the historical underpinnings of human fertility issues.

As humans, we are uniquely prone to aches and pains because of our evolutionary history. In particular, our switch to bipedalism and the development of our giant brains have forced anatomical compromise and jury-rigging throughout the human body from head to toe. From teeth with braces to feet with orthotics, we spend our lives covering up and working around anatomical shortcomings that resulted from trade-offs made by our ambitious ancestors.

FLY ME TO THE MOON

It may cause us difficulties, but a large brain is fundamental to the control of fire, written language, art, religion, and countless other aspects of human life. Human culture would look entirely different if our brains had not exploded in size in the last few million years. Other animals may pass on elements of culture such as food preferences[1] and hunting techniques,[2] but we have film festivals, tooth fairies, and snowboarding competitions. No other animal can rival us in terms of cultural depth and complexity.

We have achieved a previously unforeseen degree of intelligence because our brains are large in two ways. They are both large overall and large relative to the size of our bodies. Having one of those features can make you smart but not take-over-the-world smart. Like humans, small birds have large brains relative to their size. Their walnut-sized brains are larger compared with their bodies than ours are, and they pack their brains full of neurons to a greater degree than primates.[3] Those features give them a degree of intelligence rivaling some primates. However, the overall mass of their brains is still quite small, which explains why we study them and keep them in cages and not vice versa.

To take intelligence to the next level, an animal must excel in both measures of size. The brain needs to be both physically large and large relative to the rest of the body. Once a year I invite a couple of classrooms of local fifth graders over to the college to have a lesson about the nervous system. We watch some visual illusions, talk about stroke symptoms, and dissect sheep brains. The first thing the kids note when they have sheep brains placed in their trays is how small they are. There is a good reason sheep never make a top 10 list of the world's smartest animals. Sheep obviously have bigger brains than small birds, but their brains are relatively small compared with their overall size. What this means is that sheep are using most of their brains for controlling their vital processes (keeping the lights on, if you will) and probably not for contemplating their place in the universe. Sheep are not without their own set of skills. The part of their brains dedicated to olfaction trumps ours in both size and relative size. So they may come across as dumb to us, but we surely come across as nose dumb to them.

The brain-to-body-mass ratio explains why animals with some of the largest brains in the world do not rival us for intelligence. Elephant and whale brains are physically much larger than ours, but compared with their incredible masses their brains are relatively small. Their large brains are necessary because it takes a giant nervous system to run a giant animal. All of their anatomy is enormous. A blue whale's tongue weighs three tons! It must take an incredible number of neurons just to process the information from their tongue, let alone deal with the sensory input from the rest of their gargantuan body. Whales and elephants are smart mammals, which means that like with small birds, excelling in one of the two measures of size must help them to some degree. However, just like having a giant tongue does not necessarily make you any better at tasting, having a physically giant brain does not necessarily guarantee brilliance.

Humans took intelligence to a different level by combining those elements. The hominids that descended from trees already had decent-sized brains compared with their overall masses. Then the brain tripled in size heading down the evolutionary path that led to humans. That tripling left everybody else in the dust. It is the reason we write poetry and create governments and build bridges and fly to the moon.

Smart

Not-so-smart

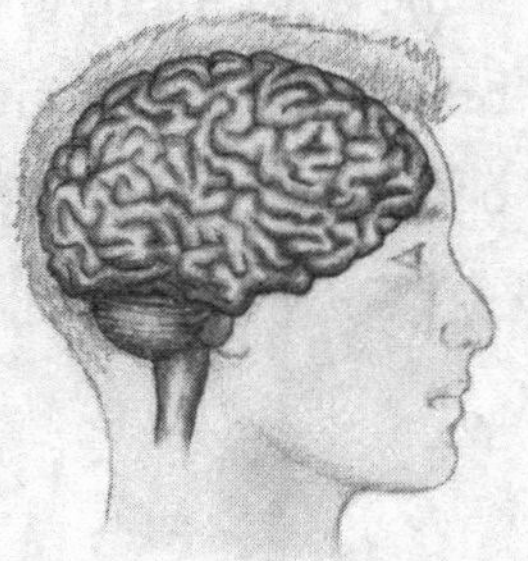

Take-over-the-world smart

BUILD THE ROCKET

Our large noodles are not the only thing that makes us human. Bipedalism is also elemental to human existence. For our ancestors, it proved to be a very energetically efficient way to travel long distances, which was important when hominins had to cover significant ground to find food. We also eventually put our freed forelimbs to skillful use. Agriculture, domestication of animals, control of fire, art, writing, carpentry—none of these complex behaviors is possible without our hands.

Our dependence on our hands remains just as true today. With just our brains, we can only dream of flying to the moon. We need our hands to turn it into reality. It is through our hands that we can truly allow our incredible minds to shine. It is interesting to note that walking upright, and the consequential freeing of the hands, happened before the increase in cranial capacity. When Lucy and her Australopithecine friends were wandering around on two feet some 3 to 4 million years ago, they were doing so with humble brains compared with our current version. It wasn't until the hands became free that the mind took off.

MORE TO THE STORY

Our oversize brains and bipedal nature have, for good reason, garnered the bulk of attention for explaining the roots of so many human anatomical maladies. Those two features are not, however, the only causes of our suffering. The shortcomings of our eyes, for example, have different origins.

The vertebrate eye evolved in the water and will always be a bit of an imperfect structure as a result. In the hundreds of millions of years since the first tetrapods crawled out of the water, eyes have come a long way. We've gone from what had to be a very blurry view for the first terrestrial vertebrates to a rather crisp image these days. Or at least, it was crisp until recently. It appears we have ourselves to blame for the new great threat to visual acuity. Our eyes are under attack by our modern, indoor lives. Recent evidence suggests eyes need a healthy amount of natural light to develop to the correct length. In the case of our eyes, modern lifestyle choices may play an even bigger role than our evolutionary history in driving our difficulties.

FORAGING FOR CHIPS AND SALSA

Our vision serves as a nice example of the power, and the limitation, of evolution. The visual acuity of animals slowly improved for hundreds of millions of years after the first vertebrates crawled out of the water. This improvement was based on two fundamental mechanisms of evolution: mutation and natural selection. As mutations arose that made for slight improvements in vision, animals with those mutations would have been at a distinct advantage. There would also have been strong selection against any mutations that made vision worse. Hundreds of millions of years later, visual acuity was still critically important to the survival and reproductive success of early humans. Hominins with poor eyesight could not hunt or forage well and without those skills had little to no shot at securing a mate. Most blurry-vision genes would have gone to the grave with their unlucky carriers.

The effects of natural selection on our ancestors extended well beyond vision. The ancient lifestyle tested their anatomy from head to toe. Hunting and foraging meant miles and miles of walking and running. Early hominins who could travel all day and not have their joints break down were the best hunters. A badly sprained ankle or a blown-out knee might have been a death sentence for any hominin in a hunter-gatherer society. Women had added anatomical pressures on top of those applied to both sexes. The increasingly large human brain intensified the difficulties of giving birth. Many women most certainly died during childbirth. Others had bodies that made them more successful at carrying, birthing, and rearing children. The shape of humans continued to evolve.

And these days? How many of us choose our mates based on traditional human skills and traits that portended success for our ancestors? I won't speak for you, but I was not overly concerned about how well women could hunt, forage, carry water, and control fire when I was dating. I am not saying natural selection is no longer at work in humans. Natural selection still affects human populations; it just affects them in different ways now than it did historically. We are no longer selecting for the same traits. In terms of survival and reproduction, it does not matter if someone has a bum knee or blurry vision these days. They can wear a knee brace and might even look cuter wearing glasses.

Instead of choosing mates based on traits and skills related to survival, we make our decisions based on conversations over margaritas at happy hour. We still choose mates based on similar global ideas (resources, security, physical attraction), but the way in which we measure those variables has changed drastically. Hunting and gathering requires a vastly separate set of skills than working at a tech start-up with great benefits. In their own times, they are both sexy career choices, but they focus selective pressures on wildly different traits.

The modern situation isn't any worse or any better. It is just different. The result is we will likely be stuck with our aches and pains for the long haul. Natural selection is not going to smooth the edges of our anatomical imperfections if those features play little to no role in who lives and who dies and how we choose our mates.

In addition, there is no stopping new mutations from cropping up in the human genome. Some mutations do lead to improvements, but most genetic mutations either cause no effect or have a deleterious outcome. Whereas natural selection may have previously filtered out mutations with a negative impact on our anatomical features, in the more modern scenario many of those mutations persist within the genome.

THE SILVER LINING

The news is not all doom and gloom. Our giant brains largely got us into this mess, and they will also help us get through it. We are already quite capable of correcting many anatomical shortcomings. We can take someone with 20/200 vision and make them 20/20. We can straighten the most crooked teeth and surgically repair damaged joints. There are a variety of contraceptive options for women who wish to reduce the frequency of menstruation. Many couples who would not have been able to reproduce in prior generations are able to have children through advances in fertility treatments.

We are also only at the dawn of a medical technological revolution that will help us live longer and more comfortable lives. Consider the techniques used to repair torn knee ligaments. Someday surgeons will look back and consider the current methods barbaric. They will be shocked when they read stories about how in the early part of this century orthopedists were still harvesting tissues from their own patients to use in surgery. Once we can grow new, perfectly matched tissues in a lab, the earlier techniques will seem as outdated as bloodletting with leeches. That day is coming, and it is probably not even very far away.

Advances in stem-cell therapy will give us new techniques in combating problems of mutation accumulation like cancer and dementia. Gene-editing tools like CRISPR will not be a panacea, but it and other breakthroughs will provide an unforeseen degree of flexibility in fixing human genetic conditions. Our ability to repair spines, retinas, feet, and all the hotspots of the body will only continue to improve in this century as we come to understand the genetic and molecular underpinnings of how a human body develops and functions. At the end of the day, however, we have a disproportionately large brain and feet that, not so far in our past, used to look more like hands. To some degree, we will never fully escape the problems driven by our unique history.

WHY US?

Since our large brains and bipedal ways are at the core of both our being and our aches and pains, it is worth considering why those features ever evolved in the first place. The great transition away from a life in the trees was the impetus. If animals do not undergo dramatic shifts in their lifestyles, they tend to click along, relatively unchanged, for extended periods of time. There are always small tweaks, because evolution never stops grinding away, but without major upheaval the changes are minor. Sharks ate fish in the ocean 400 million years ago. Sharks eat fish in the ocean today. The characteristics that made a shark successful millions of years ago are not terribly different from the characteristics that make a shark successful nowadays. As a result, sharks have not changed much over time.

Our lineage experienced significant upheaval when something drove our ancestors out of the trees a few million years ago. The reasons probably related to climate change and altered food availability. When major disruption happens, there are only two directions a species can go. It will either go extinct or it will scrape by and adapt to the new conditions. If the species is lucky enough to survive, it is because there was enough diversity within the population that there were some individuals with genetic variations amenable to the new environment. When the going gets tough, most members of a population die off, but sometimes a lucky few have the right mix of traits allowing them to live in wetter, dryer, hotter, or colder conditions.

Put another way, if every fish in a pond has the exact same set of DNA, if the pond dries up, all the fish will most likely die. If the population has a diverse enough gene pool, then a few of them might be able to wriggle into the mud and hang out until the rains come back. When the rains return, those hearty, surviving fish repopulate the pond, and the population of fish has changed, forever.

Instead of going extinct, our lineage proved to be extremely adaptable. Some of us were able to wriggle into the mud, so to speak. The move to the forest floor spurred incredible change in our ancestors, and the two most fundamental human characteristics, bipedalism and extreme intelligence, emerged out of that transitional pressure cooker. As time went on hominins covered more and more ground on two feet, and the survivors were the ones smart enough to troubleshoot the challenges of a new environment. They could thrive in social groups, communicate effectively, and cooperate in ways that led to increased protection and more efficient hunting and foraging.

Most hominins did not survive the transition. Bipedalism alone was not enough to save them. With a premium placed on adaptability and intelligence, species with smaller brains and less behavioral plasticity died out. Members of earlier hominin groups, like those in the genus *Australopithecus*, were unable to keep up as the members of our genus, *Homo*, became the dominant hominins on Earth. Hominins in other genera may have been outcompeted by members of *Homo*, or they may not have been able to adapt to new twists and turns in the ever-changing environment. Intelligence must have been critically important in those few million years after the first hominins took bipedal steps, as evidenced by the tremendous increase in cranial capacity that took place among our ancestors.

What is less clear is why only one member of the *Homo* genus made it through the evolutionary gauntlet. We are the sole survivors. What happened to *Homo erectus*, *Homo habilis*, *Homo floresiensis*, and all the other members of our tribe? Every branch except for ours on the hominin tree reaches a dead end. Why? The answers go beyond brain size. Brains from other members of *Homo* were both large and relatively large. The brains of one, *Homo neanderthalensis*, were even larger than those of *Homo sapiens*.

The answers may lie in the shape of the brains rather than the size.[4] Neanderthals dedicated a much larger proportion of their brains to visual input and had a smaller cerebellum than humans. The cerebellum has, in the last generation, become recognized as an important area of the brain for cognition, and researchers believe humans may have had a greater ability to think on their feet, problem-solve, and in short, be cognitively flexible compared with

Neanderthals. Cognitive rigidity is fine until new challenges arise. New challenges require adaptability, and Neanderthals eventually met a novel situation they could not overcome. Their final failed test may very well have come from humans, with our flexible brains and penchant for establishing dominance over all other species.

A RECIPE FOR CHANGE

The human body is a fish out of water. Or maybe a more apt comparison would be to a whale or a dolphin *in* the water. Let me try that again. The human body is a whale in the water. That's better. The pressures and forces of one environment (a life in the trees) shaped the nature of the hominid body for millions of years. Now, down on the ground, we have switched to a vastly different habitat. In the case of whales, their ancestors spent countless generations on land and were run-of-the-mill terrestrial hoofed mammals. Like all other mammals, they had lost their gills and had forelimbs and hind limbs.

Then, sometime around 50 million years ago, some of those early whale and dolphin ancestors got back in the water. We don't know exactly why, but I'm sure they had their reasons. I guarantee it was for either food, sex, safety, or all of the above. For a period of time there were transitional "whales" with a mix of terrestrial and seagoing anatomy. In other words, they could still maneuver around on land or in the intertidal zones but were also, with each passing generation, getting more comfortable back in the water.

There is a complete fossil record documenting each stage of the whale transition back to the sea.*

Now, all these years later, whales and dolphins are full-blown oceangoing animals. If they roll up on land it is bad news for them. The forelimbs became flippers, the hind limbs all but disappeared (they still possess some skeletal remnants of their hind limbs), and their noses gradually moved back to become blowholes, making it easier for them to get to the air they need to breathe. I am sure the fish all look at them funny whenever they have to surface.

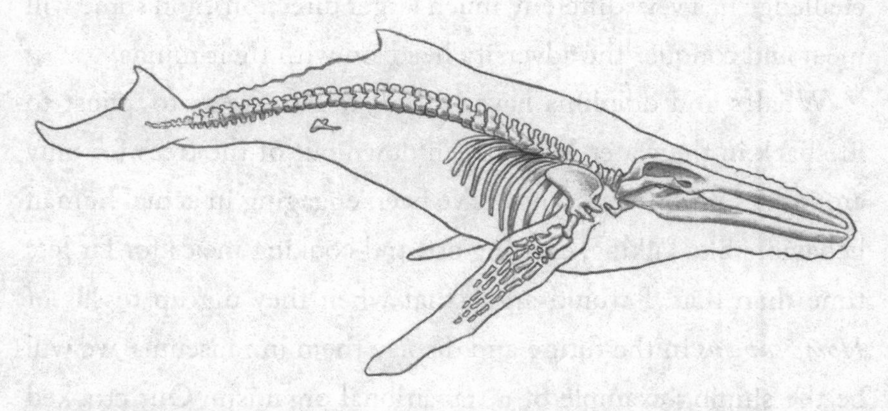

Whales still possess skeletal remnants of their hind limbs, evidence of their transition from land back to sea.

Radical environmental transitions put both whales and humans on separate evolutionary fast tracks. The challenges of drastically different lifestyles placed a premium on intelligence, and in select species of toothed whales and hominins the size of the brain increased dramatically. Both groups developed novel and efficient

---

*The University of California, Berkeley, publishes a great webpage, with references, documenting the transition.

forms of communication. Chimpanzees may be our closest relatives, but one group of toothed whales, the dolphins, are our closest rivals in terms of intelligence. Among mammals, humans have the largest brain-to-body-mass ratio. Dolphins come in second place, and no other mammal is even close.[5*] Humans and dolphins seem to suggest that exposure to a novel environment is a critical ingredient in the evolutionary recipe for extreme intelligence. Under such conditions, some species will not meet the challenge and will go extinct, some will change in other ways (blue whales took the challenge in a very different, much larger direction), and some will meet and conquer the adversity head-on with their minds.

Whales and dolphins have had 50 million years to adjust to life back in the water. We've been down out of the trees for only around 5 million years, and we've been engaging in actual human behaviors like talking, creating art, and cooking meals for far less time than that. I would argue that when they dig up fossils of *Homo sapiens* in the future and display them in museums, we will be the shining example of a transitional organism. Our crooked teeth, twisted spines, flat feet, broken ankles, constricted birth canals, and countless other features will show how we struggled to fight through the period when we ceased to be arboreal. We can only hope that, like the hagfish, we make it through the transition.

In many ways, evolution has gone wrong for us. We have error-prone bodies, and one could argue that we have become

---

*This is the answer to the last multiple choice question. Pass your papers to the left. Here are all the answers: introduction (b); 1 (c); 2 (a); 3 (d); 4 (b); 5 (a); 6 (c); 7 (c); 8 (a); 9 (b); conclusion (d).

too smart for our own good. It is entirely possible we will end up being victims of our own success. Sitting here, however, with a roof over my head, a glass of clean water on the desk, and my lunch waiting for me in the refrigerator, I can't help but appreciate the creature comforts of modern human life. Even though our large brains and bipedal ways undoubtedly cause us discomfort, I still say we came out ahead. Then again, I'm not the one who had to give birth.

Acknowledgments

I am so lucky to get to share my life and my passion for biology with my wife and daughter. Thank you for the limitless curiosity and patience you have shown with all the stories I've brought to the dinner table over the years.

Thank you to Peter Davidson for the incredible illustrations. Your creativity and talent never cease to amaze me. I smile every time I look at that bearded dragon next to the cat.

Thank you to my agent, Rick Broadhead, for taking a chance on me. *Mastodon Stew* (the original title of the book) would be confined to a folder on a hard drive if not for you. I am also indebted to all the wonderful people I've come to know at HarperCollins US and Canada. In particular, I am grateful to Julia McDowell, Peter Joseph, Brad Wilson, Jim Gifford, Natalie Meditsky, and Patricia MacDonald for shepherding a first-time author through the dense thicket of drafts and out to the light on the other side.

Joe Roberts, Tom Urquhart, and Julie Bezzerides read every word before anyone else and helped shape my writing in immeasurable ways. I cannot thank each of you enough. Thanks also to Sennett Pierce, Shane Kinzer, and Tayler Fuller for valuable feedback on individual chapters.

Thank you to all of my colleagues at Lewis-Clark State College. In particular, thanks to Eric Stoffregen and Ed Miller, who lent me an ear on countless occasions, and to Lori Stinson, who supported the project from the very beginning.

Thank you to Hailey Hulse, David Martin, Alex Helquist, Alan Hain, and Laura Bracken for sharing your stories with me.

Thank you to Dan Riskin and David Quammen for helping me navigate the publishing waters and to Amy McCune, Holly Dunsworth, Neil Shubin, and Jeremy DeSilva for your clear and prompt answers to my scientific queries.

Lastly, thank you to Mike and Sylvia, Nick and Heidi, and my parents, Ted and Betty, for all your love and support. I grew up in a house with a scientist and a writer, and this book feels like a natural outcome of that upbringing. I hope you all enjoyed reading it as much as I enjoyed writing it.

Notes

INTRODUCTION: It's Not Your Fault

1. Pauli, Jonathan N., et al. "A Syndrome of Mutualism Reinforces the Lifestyle of a Sloth." *Proceedings of the Royal Society B: Biological Sciences* 281, no. 1778 (2014): 20133006. doi:10.1098/rspb.2013.3006

CHAPTER 1: Mastodon Stew

1. Fraser, Gareth J., et al. "The Odontode Explosion: The Origin of Tooth-Like Structures in Vertebrates." *BioEssays* 32, no. 9 (2010): 808–817.

2. Johanson, Zerina, et al. "Origin and Evolution of Gnathostome Dentitions: A Question of Teeth and Pharyngeal Denticles in Placoderms." *Biological Reviews* 80, no. 2 (2005): 303–345.

3. Zhu, Min, et al. "A Silurian Maxillate Placoderm Illuminates Jaw Evolution." *Science* 354, no. 6310 (2016): 334–336.

4. Knapp, Landon, et al. "Conservation Status of the World's Hagfish Species and the Loss of Phylogenetic Diversity and Ecosystem Function." *Aquatic Conservation: Marine and Freshwater Ecosystems* 21, no. 5 (2011): 401–411.

5. Paganini-Hill, Annlia, et al. "Dental Health Behaviors, Dentition, and Mortality in the Elderly: The Leisure World Cohort Study." *Journal of Aging Research* (2011): 2011156061. doi:10.4061/2011/256601

6. Swindler, Daris. *Primate Dentition: An Introduction to the Teeth of Non-Human Primates*. Cambridge: Cambridge University Press, 2002.

7. Hammond, Ashley S., et al. "Middle Miocene *Pierolapithecus* Provides a First Glimpse into Early Hominid Pelvic Morphology." *Journal of Human Evolution* 64, no. 6 (2013): 658–666.

8. Macho, Gabriele A., et al. "An Exploratory Study on the Combined Effects of External and Internal Morphology on Load Dissipation in Primate Capitates: Its Potential for an Understanding of the Positional and Locomotor Repertoire of Early Hominins." *Folia Primatologica* 81, no. 5 (2010): 292–304.

9. Sponheimer, Matt, et al. "Isotopic Evidence of Early Hominin Diets." *Proceedings of the National Academy of Sciences* 110, no. 26 (2013): 10513–10518.

10. Domínguez-Rodrigo, Manuel, et al. "Cutmarked Bones from Pliocene Archaeological Sites at Gona, Afar, Ethiopia: Implications for the Function of the World's Oldest Stone Tools." *Journal of Human Evolution* 48, no. 2 (2005): 109–121.

11. Stedman, Hansell H., et al. "Myosin Gene Mutation Correlates with Anatomical Changes in the Human Lineage." *Nature* 428, no. 6981 (2004): 415–418.

12. Berna, Francesco, et al. "Microstratigraphic Evidence of In Situ Fire in the Acheulean Strata of Wonderwerk Cave, Northern Cape Province, South Africa." *Proceedings of the National Academy of Sciences* 109, no. 20 (2012): E1215–E1220. doi:10.1073/pnas.1117620109

13. Shimelmitz, Ron, et al. "'Fire at Will': The Emergence of Habitual Fire Use 350,000 Years Ago." *Journal of Human Evolution* 77 (2014): 196–203.

14. Sandgathe, Dennis M., et al. "Timing of the Appearance of Habitual Fire Use." *Proceedings of the National Academy of Sciences* 108, no. 29 (2011): E298. doi:10.1073/pnas.1106759108

15. Smith, Alex R., et al. "The Significance of Cooking for Early Hominin Scavenging." *Journal of Human Evolution* 84 (2015): 62–70.

16. Carmody, Rachel N., et al. "Energetic Consequences of Thermal and Non-thermal Food Processing." *Proceedings of the National Academy of Sciences* 108, no. 48 (2011): 19199–19203.

17. Wobber, Victoria, et al. "Great Apes Prefer Cooked Food." *Journal of Human Evolution* 55, no. 2 (2008): 340–348.

18. Wu, Xiaohong, et al. "Early Pottery at 20,000 Years Ago in Xianrendong Cave, China." *Science* 336, no. 6089 (2012): 1696–1700.

19. Snir, Ainit, et al. "The Origin of Cultivation and Proto-Weeds, Long before Neolithic Farming." *PLOS ONE* 10, no. 7 (2015): e0131422. doi:10.1371/journal.pone.0131422

20. Pinhasi, Ron, et al. "Incongruity between Affinity Patterns Based on Mandibular and Lower Dental Dimensions Following the Transition to Agriculture in the Near East, Anatolia and Europe." *PLOS ONE* 10, no. 2 (2015): e0117301. doi.org/10.1371/journal.pone.0117301

21. Brace, Loring C., et al. "Gradual Change in Human Tooth Size in the Late Pleistocene and Post-Pleistocene." *Evolution* 41, no. 4 (1987): 705–720.

22. Lieberman, Daniel E., et al. "Effects of Food Processing on Masticatory Strain and Craniofacial Growth in a Retrognathic Face." *Journal of Human Evolution* 46, no. 6 (2004): 655–677.

23. Lyons, Kathleen S., et al. "Of Mice, Mastodons and Men: Human-Mediated Extinctions on Four Continents." *Evolutionary Ecology Research* 6, no. 3 (2004): 339–358.

CHAPTER 2: The Fish-Eye Lens

1. Williams, Katie M., et al. "Prevalence of Refractive Error in Europe: The European Eye Epidemiology (E^3) Consortium." *European Journal of Epidemiology* 30, no. 4 (2015): 305–315.

2. Lamb, Trevor D., et al. "Evolution of the Vertebrate Eye: Opsins, Photoreceptors, Retina and Eye Cup." *Nature Reviews Neuroscience* 8, no. 12 (2007): 960–976.

3. Romer, Alfred S. "Tetrapod Limbs and Early Tetrapod Life." *Evolution* 12, no. 3 (1958): 365–369.

4. Retallack, Gregory J. "Woodland Hypothesis for Devonian Tetrapod Evolution." *Journal of Geology* 119, no. 3 (2011): 235–258.

5. Nakano, Tamami, et al. "Blink-Related Momentary Activation of the Default Mode Network While Viewing Videos." *Proceedings of the National Academy of Sciences* 110, no. 2 (2013): 702–706.

6. Bowmaker, James K. "Evolution of Vertebrate Visual Pigments." *Vision Research* 48, no. 20 (2008): 2022–2041.

7. Angielczyk, K.D., et al. "Nocturnality in Synapsids Predates the Origin of Mammals by over 100 Million Years." *Proceedings of the Royal Society of London B: Biological Sciences* 281, no. 1793 (2014): 20141642. doi:10.1098/rspb.2014.1642

8. Jordan, Gabriele, et al. "The Dimensionality of Color Vision in Carriers of Anomalous Trichromacy." *Journal of Vision* 10, no. 8 (2010): 1–19.

9. Eisner, Thomas, et al. "Plant Taxonomy: Ultraviolet Patterns of Flowers Visible as Fluorescent Patterns in Pressed Herbarium Specimens." *Science* 179, no. 4072 (1973): 486–487.

10. Gronquist, Matthew, et al. "Attractive and Defensive Functions of the Ultraviolet Pigments of a Flower (*Hypericum calycinum*)." *Proceedings of the National Academy of Sciences* 98, no. 24 (2001): 13745–13750.

11. Douglas, R.H., et al. "The Spectral Transmission of Ocular Media Suggests Ultraviolet Sensitivity Is Widespread among Mammals." *Proceedings of the Royal Society of London B: Biological Sciences* 281, no. 1780 (2014): 20132995. doi:10.1098/rspb.2013.2995

12. Collier, R., et al. "The Gray Squirrel Lens Protects the Retina from Near-UV Radiation Damage." *Progress in Clinical and Biological Research* 247 (1987): 571–585.

13. Dolgin, Elie. "The Myopia Boom." *Nature* 519, no. 7543 (2015): 276–278.

14. Jones, Lisa A., et al. "Parental History of Myopia, Sports and Outdoor Activities, and Future Myopia." *Investigative Ophthalmology & Visual Science* 48, no. 8 (2007): 3524–3532.

15. Rose, Kathryn A., et al. "Outdoor Activity Reduces the Prevalence of Myopia in Children." *Ophthalmology* 115, no. 8 (2008): 1279–1285.

CHAPTER 3: **Down the Hatch**

1. Cyr, Claude, Canadian Paediatric Society, Injury Prevention Committee. "Preventing Choking and Suffocation in Children." *Paediatrics & Child Health* 17, no. 2 (2012): 91–92.

2. Griscom, N.T., et al. "Dimensions of the Growing Trachea Related to Age and Gender." *American Journal of Roentgenology* 146, no. 2 (1986): 233–237.

3. American Academy of Pediatrics. "Policy Statement—Prevention of Choking among Children." *Pediatrics* 125, no. 3 (2010): 601–607.

4. Ekberg, Olle, et al. "Clinical and Demographic Data in 75 Patients with Near-Fatal Choking Episodes." *Dysphagia* 7, no. 4 (1992): 205–208.

5. Longo, Sarah, et al. "Homology of Lungs and Gas Bladders: Insights from Arterial Vasculature." *Journal of Morphology* 274, no. 6 (2013): 687–703.

6. Weinberg, Samantha. *A Fish Caught in Time.* New York: HarperCollins, 2000.

7. Cupello, Camila, et al. "Allometric Growth in the Extant Coelacanth Lung during Ontogenetic Development." *Nature Communications* 6, no. 8222 (2015): 1–5.

8. Biro, Peter. "The Evolutionary Reason for the Need to Secure the Airway in Anesthesiology." *Israel Medical Association Journal* 13, no. 1 (2011): 5–8.

9 Saigusa, Hideto. "Comparative Anatomy of the Larynx and Related Structures." *Japan Medical Association* 54 (2011): 241–247.

10. Kotpal, R.L. *Modern Textbook of Zoology: Vertebrates.* Meerut, India: Rastogi Publications, 2010.

11. Laitman, Jeffrey T., et al. "Specializations of the Human Upper Respiratory and Upper Digestive Systems as Seen through Comparative and Developmental Anatomy." *Dysphagia* 8, no. 4 (1993): 318–325.

12. Beckers, Gabriël J.L., et al. "Vocal-Tract Filtering by Lingual Articulation in a Parrot." *Current Biology* 14, no. 17 (2004): 1592–1597.

13. Lieberman, Philip, et al. "Tracking the Evolution of Language and Speech: Comparing Vocal Tracts to Identify Speech Capabilities." *Expedition: The Magazine of the University of Pennsylvania* 49, no. 2 (2007): 15–20.

14. Hublin, Jean-Jacques, et al. "New Fossils from Jebel Irhoud, Morocco and the Pan-African Origin of *Homo sapiens.*" *Nature* 546, no. 7657 (2017): 289–292.

15. Lieberman, Philip, et al. "The Evolution of Human Speech: Its Anatomical and Neural Bases." *Current Anthropology* 48, no. 1 (2007): 39–66.

16. Higham, Tom, et al. "The Timing and Spatiotemporal Patterning of Neanderthal Disappearance." *Nature* 512, no. 7514 (2014): 306–309.

17. Young, Terry, et al. "The Occurrence of Sleep-Disordered Breathing among Middle-Aged Adults." *New England Journal of Medicine* 328, no. 17 (1993): 1230–1235.

18. Davidson, Terence M. "The Great Leap Forward: The Anatomic Basis for the Acquisition of Speech and Obstructive Sleep Apnea." *Sleep Medicine* 4, no. 3 (2003): 185–194.

19. Ramaihgari, Bhavitha, et al. "Zinc Nanoparticles Enhance Brain Connectivity in the Canine Olfactory Network: Evidence from an fMRI Study in Unrestrained Awake Dogs." *Frontiers in Veterinary Science* 5 (2018). doi:10.3389/fvets.2018.00127

CHAPTER 4: **Two Snakes a Day**

1. Stahel, Philip F., et al. "Wrong-Site and Wrong-Patient Procedures in the Universal Protocol Era: Analysis of a Prospective Database of Physician Self-Reported Occurrences." *Archives of Surgery* 145, no. 10 (2010): 978–984.

2. Drosos, G.I., et al. "The Causes and Mechanisms of Meniscal Injuries in the Sporting and Non-Sporting Environment in an Unselected Population." *The Knee* 11, no. 2 (2004): 143–149.

3. Drosos, G.I., et al. "The Causes and Mechanisms of Meniscal Injuries," 143–149.

4. Raichlen, David A., et al. "Laetoli Footprints Preserve Earliest Direct Evidence of Human-Like Bipedal Biomechanics." *PLOS ONE* 5, no. 3 (2010): e9769. doi:10.1371/journal.pone.0009769

5. Niemitz, Carsten. "The Evolution of the Upright Posture and Gait—A Review and a New Synthesis." *Naturwissenschaften* 97, no. 3 (2010): 241–263.

6. WoldeGabriel, Giday, et al. "Geology and Palaeontology of the Late Miocene Middle Awash Valley, Afar Rift, Ethiopia." *Nature* 412, no. 6843 (2001): 175–178.

7. Cuthbert, Mark O., et al. "A Spring Forward for Hominin Evolution in East Africa." *PLOS ONE* 9, no. 9 (2014): e107358. doi:10.1371/journal.pone.0107358

8. Niemitz, Carsten. "The Evolution of the Upright Posture and Gait," 241–263.

9. Lovejoy, C. Owen. "Reexamining Human Origins in Light of *Ardipithecus ramidus*." *Science* 326, no. 5949 (2009): 74e1–74e8. doi:10.1126/science.1175834

10. Sockol, Michael D., et al. "Chimpanzee Locomotor Energetics and the Origin of Human Bipedalism." *Proceedings of the National Academy of Sciences* 104, no. 30 (2007): 12265–12269.

11. Lebel, B., et al. "Ontogeny-Phylogeny." In *The Meniscus*, edited by Phillippe Beaufils and René Verdonk, 3–9. Berlin Heidelberg: Springer-Verlag, 2010.

12. Shad, Jimmy, et al. "Rare Disease: An Infant with Caudal Appendage." *BMJ Case Reports* (2012): bcr1120115160. doi:10.1136/bcr.11.2011.5160

13. Schmidt, H. "Supernumerary Nipples: Prevalence, Size, Sex and Side Predilection—A Prospective Clinical Study." *European Journal of Pediatrics* 157, no. 10 (1998): 821–823.

14. Jordan, Michael R. "Lateral Meniscal Variants: Evaluation and Treatment." *Journal of the American Academy of Orthopaedic Surgeons* 4, no. 4 (1996): 191–200.

15. Yaniv, Moshe, et al. "The Discoid Meniscus." *Journal of Children's Orthopaedics* 1, no. 2 (2007): 89–96.

16. Lebel, B., et al. "Ontogeny-Phylogeny," 3–9.

17. Flouzat-Lachaniette, et al. "Discoid Medial Meniscus: Report of Four Cases and Literature Review." *Orthopaedics & Traumatology: Surgery & Research* 97, no. 8 (2011): 826–832.

18. Sutton, Karen M., et al. "Anterior Cruciate Ligament Rupture: Differences between Males and Females." *Journal of the American Academy of Orthopaedic Surgeons* 21, no. 1 (2013): 41–50.

19. Prodromos, Chadwick C., et al. "A Meta-Analysis of the Incidence of Anterior Cruciate Ligament Tears as a Function of Gender, Sport, and a Knee Injury–Reduction Regimen." *Arthroscopy: The Journal of Arthroscopic & Related Surgery* 23, no. 12 (2007): 1320-1325.

20. Ireland, Mary Lloyd. "The Female ACL: Why Is It More Prone to Injury?" *Orthopedic Clinics* 33, no. 4 (2002): 637–651.

21. Giugliano, Danica N., et al. "ACL Tears in Female Athletes." *Physical Medicine and Rehabilitation Clinics of North America* 18, no. 3 (2007): 417–438.

22. Chaudhari, Ajit M.W., et al. "Anterior Cruciate Ligament-Injured Subjects Have Smaller Anterior Cruciate Ligaments than Matched Controls: A Magnetic Resonance Imaging Study." *American Journal of Sports Medicine* 37, no. 7 (2009): 1282–1287.

23. Voskanian, Natalie. "ACL Injury Prevention in Female Athletes: Review of the Literature and Practical Considerations in Implementing an ACL Prevention Program." *Current Reviews in Musculoskeletal Medicine* 6, no. 2 (2013): 158–163.

CHAPTER 5: **The Honey Holiday**

1. Venkataraman, Vivek V., et al. "Tree Climbing and Human Evolution." *Proceedings of the National Academy of Sciences* 110, no. 4 (2013): 1237–1242.

2. Harako, R. "Ecological and Sociological Importance of Honey to the Mbuti Net Hunters Eastern Zaire, Kyoto." *African Study Monographs* 1 (1981): 55–68.

3. Bailey, Robert Converse. *The Behavioral Ecology of Efe Pygmy Men in the Ituri Forest, Zaire*. No. 86. University of Michigan Museum, 1991.

4. Risser, Daniele, et al. "Risk of Dying after a Free Fall from Height." *Forensic Science International* 78, no. 3 (1996): 187–191.

5. White, Tim D., et al. "*Ardipithecus ramidus* and the Paleobiology of Early Hominids." *Science* 326, no. 5949 (2009): 64–86.

6. Nix, Sheree, et al. "Prevalence of Hallux Valgus in the General Population: A Systematic Review and Meta-Analysis." *Journal of Foot and Ankle Research* 3, no. 21 (2010). doi.org/10.1186/1757-1146-3-21

7. Demenocal, Peter B. "African Climate Change and Faunal Evolution during the Pliocene–Pleistocene." *Earth and Planetary Science Letters* 220, no. 1–2 (2004): 3–24.

8. Bramble, Dennis M., et al. "Endurance Running and the Evolution of *Homo*." *Nature* 432, no. 7015 (2004): 345–352.

9. Cunningham, C.B., et al. "The Influence of Foot Posture on the Cost of Transport in Humans." *Journal of Experimental Biology* 213, no. 5 (2010): 790–797.

10. Lieberman, Daniel E., et al. "Foot Strike Patterns and Collision Forces in Habitually Barefoot Versus Shod Runners." *Nature* 463, no. 7280 (2010): 531–535.

11. Godin, Alfred J. *Wild Mammals of New England*. Baltimore: Johns Hopkins University Press, 1977.

12. Cheung, Jason Tak-Man, et al. "Consequences of Partial and Total Plantar Fascia Release: A Finite Element Study." *Foot & Ankle International* 27, no. 2 (2006): 125–132.

13. DeSilva, Jeremy M., et al. "Lucy's Flat Feet: The Relationship between the Ankle and Rearfoot Arching in Early Hominins." *PLOS ONE* 5, no. 12 (2010): e14432. doi:10.1371/journal.pone.0014432

14. Young, Richard W. "Evolution of the Human Hand: The Role of Throwing and Clubbing." *Journal of Anatomy* 202, no. 1 (2003): 165–174.

15. Persons, W. Scott, et al. "The Functional Origin of Dinosaur Bipedalism: Cumulative Evidence from Bipedally Inclined Reptiles and Disinclined Mammals." *Journal of Theoretical Biology* 420 (2017): 1–7.

16. Turner, Alan H., et al. "Feather Quill Knobs in the Dinosaur Velociraptor." *Science* 317, no. 5845 (2007): 1721.

CHAPTER 6: **Baby Got Back . . . Pain**

1. Rubin, Devon I. "Epidemiology and Risk Factors for Spine Pain." *Neurologic Clinics* 25, no. 2 (2007): 353–371.

2. Vos, Theo, et al. "Global, Regional, and National Incidence, Prevalence, and Years Lived with Disability for 301 Acute and Chronic Diseases and Injuries in 188 Countries, 1990–2013: A Systematic Analysis for the Global Burden of Disease Study 2013." *The Lancet* 386, no. 9995 (2015): 743–800.

3. Krebs, Erin E., et al. "Effect of Opioid vs Nonopioid Medications on Pain-Related Function in Patients with Chronic Back Pain or Hip or Knee Osteoarthritis Pain: The SPACE Randomized Clinical Trial." *Journal of the American Medical Association* 319, no. 9 (2018): 872–882.

4. Hedegaard, Holly, et al. "Drug Overdose Deaths in the United States, 1999–2015." *NCHS Data Brief*, no. 273 (2017): 1–8.

5. Lewis, David M.G., et al. "Lumbar Curvature: A Previously Undiscovered Standard of Attractiveness." *Evolution and Human Behavior* 36, no. 5 (2015): 345–350.

6. Whitcome, Katherine K., et al. "Fetal Load and the Evolution of Lumbar Lordosis in Bipedal Hominins." *Nature* 450, no. 7172 (2007): 1075–1078.

7. Lin, R.M., et al. "Lumbar Lordosis: Normal Adults." *Journal of the Formosan Medical Association* 91, no. 3 (1992): 329–333.

8. Fernand, Robert, et al. "Evaluation of Lumbar Lordosis. A Prospective and Retrospective Study." *Spine* 10, no. 9 (1985): 799–803.

9. Lewis, David M.G., et al. "Lumbar Curvature," 345–350.

10. Plomp, Kimberly A., et al. "The Ancestral Shape Hypothesis: An Evolutionary Explanation for the Occurrence of Intervertebral Disc Herniation in Humans." *BMC Evolutionary Biology* 15, no. 68 (2015). doi.org/10.1186/s12862-015-0336-y

11. Plomp, Kimberly A., et al. "The Ancestral Shape Hypothesis."

12. Risbud, Makarand V., et al. "Notochordal Cells in the Adult Intervertebral Disc: New Perspective on an Old Question." *Critical Reviews™ in Eukaryotic Gene Expression* 21, no. 1 (2011): 29–41.

13. Hejnol, Andreas, et al. "Animal Evolution: Stiff or Squishy Notochord Origins?" *Current Biology* 24, no. 23 (2014): 1131–1133.

14. Jacob, Tamar, et al. "Physical Activities and Low Back Pain: A Community-Based Study." *Medicine & Science in Sports & Exercise* 36, no. 1 (2004): 9–15.

15. Dijken, Christina Björck-van, et al. "Low Back Pain, Lifestyle Factors and Physical Activity: A Population-Based Study." *Journal of Rehabilitation Medicine* 40, no. 10 (2008): 864–869.

16. Heneweer, Hans, et al. "Physical Activity and Low Back Pain: A U-Shaped Relation?" *Pain* 143, no. 1–2 (2009): 21–25.

17. Kujala, Urho M., et al. "Physical Loading and Performance as Predictors of Back Pain in Healthy Adults: A 5-Year Prospective Study." *European Journal of Applied Physiology and Occupational Physiology* 73, no. 5 (1996): 452–458.

18. Hartvigsen, Jan, et al. "Is Sitting-While-at-Work Associated with Low Back Pain? A Systematic, Critical Literature Review." *Scandinavian Journal of Public Health* 28, no. 3 (2000): 230–239.

19. Smith, Peter, et al. "The Relationship between Occupational Standing and Sitting and Incident Heart Disease over a 12-Year Period in Ontario, Canada." *American Journal of Epidemiology* 187, no. 1 (2017): 27–33.

20. Samson, David R., et al. "Chimpanzees Preferentially Select Sleeping Platform Construction Tree Species with Biomechanical Properties That Yield Stable, Firm, but Compliant Nests." *PLOS ONE* 9, no. 4 (2014): e95361. doi.org/10.1371/journal.pone.0095361

21. Samson, David R., et al. "Chimpanzees."

22. Wadley, Lyn, et al. "Middle Stone Age Bedding Construction and Settlement Patterns at Sibudu, South Africa." *Science* 334, no. 6061 (2011): 1388–1391.

23. Samson, David R., et al. "Sleep Intensity and the Evolution of Human Cognition." *Evolutionary Anthropology: Issues, News, and Reviews* 24, no. 6 (2015): 225–237.

24. Levy, H., et al. "Mattresses and Sleep for Patients with Low Back Pain: A Survey of Orthopaedic Surgeons." *Journal of the Southern Orthopaedic Association* 5, no. 3 (1996): 185–187.

25. Kovacs, Francisco M., et al. "Effect of Firmness of Mattress on Chronic Non-Specific Low-Back Pain: Randomised, Double-Blind, Controlled, Multicentre Trial." *The Lancet* 362, no. 9396 (2003): 1599–1604.

CHAPTER 7: **To Bleed or Not to Bleed**

1. Dasharathy, Sonya S., et al. "Menstrual Bleeding Patterns among Regularly Menstruating Women." *American Journal of Epidemiology* 175, no. 6 (2012): 536–545.

2. Chiazze, Leonard, et al. "The Length and Variability of the Human Menstrual Cycle." *Journal of the American Medical Association* 203, no. 6 (1968): 377–380.

3. Vos, Theo, et al. "Global, Regional, and National Incidence, Prevalence, and Years Lived with Disability for 301 Acute and Chronic Diseases and Injuries in 188 Countries, 1990–2013: A Systematic Analysis for the Global Burden of Disease Study 2013." *The Lancet* 386, no. 9995 (2015): 743–800.

4. Wilkins, L., et al. "Macrogenitosomia Precox Associated with Hyperplasia of the Androgenic Tissue of the Adrenal and Death from Corticoadrenal Insufficiency Case Report." *Endocrinology* 26, no. 3 (1940): 385–395.

5. Khan, Yasir, et al. "Pica in Iron Deficiency: A Case Series." *Journal of Medical Case Reports* 4, no. 86 (2010). doi.org/10.1186/1752-1947-4-86

6. Milman, Nils. "Serum Ferritin in Danes: Studies of Iron Status from Infancy to Old Age, During Blood Donation and Pregnancy." *International Journal of Hematology* 63, no. 2 (1996): 103–135.

7. Wilson, Don E., et al. *Mammal Species of the World: A Taxonomic and Geographic Reference*. Baltimore: Johns Hopkins University Press, 2005.

8. Strassmann, Beverly I. "The Biology of Menstruation in *Homo sapiens*: Total Lifetime Menses, Fecundity, and Nonsynchrony in a Natural-Fertility Population." *Current Anthropology* 38, no. 1 (1997): 123–129.

9. Strassmann, Beverly I. "The Biology of Menstruation," 123–129.

10. Rees, M. "The Age of Menarche." *ORGYN* 4 (1995): 2–4.

11. Howdeshell, Kembra L., et al. "Environmental Toxins: Exposure to Bisphenol A Advances Puberty." *Nature* 401, no. 6755 (1999): 763–764.

12. Lev. 15:19–33.

13. Profet, Margie. "Menstruation as a Defense against Pathogens Transported by Sperm." *Quarterly Review of Biology* 68, no. 3 (1993): 335–386.

14. Strassmann, Beverly I. "Energy Economy in the Evolution of Menstruation." *Evolutionary Anthropology* 5, no. 5 (1996): 157–164.

15. Zhu, Ha, et al. "Endometrial Stromal Cells and Decidualized Stromal Cells: Origins, Transformation and Functions." *Gene* 551, no. 1 (2014): 1–14.

16. Xu, X.B., et al. "Menstrual-Like Changes in Mice Are Provoked through the Pharmacologic Withdrawal of Progesterone Using Mifepristone following Induction of Decidualization." *Human Reproduction* 22, no. 12 (2007): 3184–3191.

17. Emera, Deena, et al. "The Evolution of Menstruation: A New Model for Genetic Assimilation." *BioEssays* 34, no. 1 (2012): 26–35.

18. Von Rango, U., et al. "Apoptosis of Extravillous Trophoblast Cells Limits the Trophoblast Invasion in Uterine but Not in Tubal Pregnancy during First Trimester." *Placenta* 24, no. 10 (2003): 929–940.

19. Martin, Robert D. "Human Reproduction: A Comparative Background for Medical Hypotheses." *Journal of Reproductive Immunology* 59, no. 2 (2003): 111–135.

20. Teklenburg, Gijs, et al. "Natural Selection of Human Embryos: Decidualizing Endometrial Stromal Cells Serve as Sensors of Embryo Quality upon Implantation." *PLOS ONE* 5, no. 4 (2010): e10258. doi.org/10.1371/journal.pone.0010258

21. Salker, Madhuri, et al. "Natural Selection of Human Embryos: Impaired Decidualization of Endometrium Disables Embryo-Maternal Interactions and Causes Recurrent Pregnancy Loss." *PLOS ONE* 5, no. 4 (2010): e10287. doi.org/10.1371/journal.pone.0010287

22. Salker, Madhuri, et al. "Natural Selection of Human Embryos."

CHAPTER 8: **Absence Makes the Heart Grow Fonder . . . and the Penis Thrust More Deeply**

1. Evers, Johannes L.H. "Female Subfertility." *The Lancet* 360, no. 9327 (2002): 151–159.

2. Gibbons, Robert Alexander. *A Lecture on Sterility: Its Etiology and Treatment, Etc.* London: J. & A. Churchill, 1911.

3. Ling, Constance M., et al. "Extrapelvic Endometriosis: A Case Report and Review of the Literature." *Journal SOGC* 22, no. 2 (2000): 97–100.

4. Lukas, Dieter, et al. "The Evolution of Infanticide by Males in Mammalian Societies." *Science* 346, no. 6211 (2014): 841–844.

5. Van Schaik, Carel P., et al. "Mating Conflict in Primates: Infanticide, Sexual Harassment and Female Sexuality." In *Sexual Selection in Primates: New and Comparative Perspectives*, edited by Peter Kappeler and Carel van Schaik, 131–150. Cambridge: Cambridge University Press, 2004.

6. Miller, Geoffrey, et al. "Ovulatory Cycle Effects on Tip Earnings by Lap Dancers: Economic Evidence for Human Estrus?" *Evolution and Human Behavior* 28, no. 6 (2007): 375–381.

7. Kuukasjärvi, Seppo, et al. "Attractiveness of Women's Body Odors over the Menstrual Cycle: The Role of Oral Contraceptives and Receiver Sex." *Behavioral Ecology* 15, no. 4 (2004): 579–584.

8. Havlíček, Jan, et al. "Non-Advertized Does Not Mean Concealed: Body Odour Changes across the Human Menstrual Cycle." *Ethology* 112, no. 1 (2006): 81–90.

9. Miller, Saul L., et al. "Scent of a Woman: Men's Testosterone Responses to Olfactory Ovulation Cues." *Psychological Science* 21, no. 2 (2010): 276–283.

10. Kirchengast, S., et al. "Changes in Fat Distribution (WHR) and Body Weight across the Menstrual Cycle." *Collegium Antropologicum* 26 (2002): 47–57.

11. Roberts, S. Craig, et al. "Female Facial Attractiveness Increases during the Fertile Phase of the Menstrual Cycle." *Proceedings of the Royal Society of London B: Biological Sciences* 271, no. 5 (2004): S270–S272. doi:10.1098/rsbl.2004.0174

12. Symonds, C.S., et al. "Effects of the Menstrual Cycle on Mood, Neurocognitive and Neuroendocrine Function in Healthy Premenopausal Women." *Psychological Medicine* 34, no. 1 (2004): 93–102.

13. Haselton, Martie G., et al. "Ovulatory Shifts in Human Female Ornamentation: Near Ovulation, Women Dress to Impress." *Hormones and Behavior* 51, no. 1 (2007): 40–45.

14. Kumar, A., et al. "Swinging High and Low: Why Do the Testes Hang at Different Levels? A Theory on Surface Area and Thermoregulation." *Medical Hypotheses* 70 (2008): 698–708.

15. Gallup Jr., Gordon G., et al. "On the Origin of Descended Scrotal Testicles: The Activation Hypothesis." *Evolutionary Psychology* 7, no. 4 (2009): 517–526.

16. Møller, Anders Pape. "Ejaculate Quality, Testes Size and Sperm Competition in Primates." *Journal of Human Evolution* 17, no. 5 (1988): 479–488.

17. Martin, P.A., et al. "The Effect of Ratios and Numbers of Spermatozoa Mixed from Two Males on Proportions of Offspring." *Reproduction* 39, no. 2 (1974): 251–258.

18. Harcourt, Alexander H., et al. "Testis Weight, Body Weight and Breeding System in Primates." *Nature* 293, no. 5827 (1981): 55–57.

19. Warner, Harold, et al. "Electroejaculation of the Great Apes." *Annals of Biomedical Engineering* 2, no. 4 (1974): 419–432.

20. Gallup Jr., Gordon G., et al. "The Human Penis as a Semen Displacement Device." *Evolution and Human Behavior* 24, no. 4 (2003): 277–289.

21. Shackelford, Todd K., et al. "Adaptation to Sperm Competition in Humans." *Current Directions in Psychological Science* 16, no. 1 (2007): 47–50.

22. Baker, R. Robin, et al. "Number of Sperm in Human Ejaculates Varies in Accordance with Sperm Competition Theory." *Animal Behaviour* 37, no. 1–2 (1989): 867–869.

23. Gallup Jr., Gordon G., et al. "The Human Penis," 277–289.

24. Shackelford, Todd K., et al. "Adaptation to Sperm Competition," 47–50.

25. Lee, Amy, et al. "Early Human Embryos Are Naturally Aneuploid—Can That Be Corrected?" *Journal of Assisted Reproduction and Genetics* 34, no. 1 (2017): 15–21.

26. Van Echten-Arends, Jannie, et al. "Chromosomal Mosaicism in Human Preimplantation Embryos: A Systematic Review." *Human Reproduction Update* 17, no. 5 (2011): 620–627.

27. Lee, Amy, et al. "Early Human Embryos," 15–21.

28. Stewart, Elizabeth A., et al. "Epidemiology of Uterine Fibroids: A Systematic Review." *BJOG: An International Journal of Obstetrics & Gynaecology* 124, no. 10 (2017): 1501–1512.

29. Levine, Hagai, et al. "Temporal Trends in Sperm Count: A Systematic Review and Meta-Regression Analysis." *Human Reproduction Update* 23, no. 6 (2017): 646–659.

CHAPTER 9: **The Greatest Pain of All**

1. Profet, Margie. "Pregnancy Sickness as Adaptation: A Deterrent to Maternal Ingestion of Teratogens." In *The Adapted Mind: Evolutionary Psychology and the Generation of Culture*, edited by J.H. Barkow, L. Cosmides, and J. Tooby, 327–366. New York: Oxford University Press, 1992.

2. Sherman, Paul W., et al. "Protecting Ourselves from Food: Spices and Morning Sickness May Shield Us from Toxins and Microorganisms in the Diet." *American Scientist* 89, no. 2 (2001): 142–151.

3. Profet, Margie. "Pregnancy Sickness as Adaptation," 327–366.

4. Weigel, Ronald M., et al. "Nausea and Vomiting of Early Pregnancy and Pregnancy Outcome: A Meta-Analytical Review." *BJOG: An International Journal of Obstetrics & Gynaecology* 96, no. 11 (1989): 1312–1318.

5. Sherman, Paul W., et al. "Protecting Ourselves from Food," 142–151.

6. Savitz, D.A., et al. "Ethnicity and Gestational Diabetes in New York City, 1995–2003." *BJOG: An International Journal of Obstetrics & Gynaecology* 115, no. 8 (2008): 969–978.

7. Brown, Elizabeth A., et al. "Many Ways to Die, One Way to Arrive: How Selection Acts Through Pregnancy." *Trends in Genetics* 29, no. 10 (2013): 585-592.

8. Jolly, Matthew C., et al. "Risk Factors for Macrosomia and Its Clinical Consequences: A Study of 350,311 Pregnancies." *European Journal of Obstetrics & Gynecology and Reproductive Biology* 111, no. 1 (2003): 9–14.

9. Robillard, Pierre-Yves, et al. "Preeclampsia/Eclampsia and the Evolution of the Human Brain." *Current Anthropology* 44, no. 1 (2003): 130–135.

10. Martin, Robert D. "Scaling of the Mammalian Brain: The Maternal Energy Hypothesis." *Physiology* 11, no. 4 (1996): 149–156.

11. Hodgins, Stephen. "Pre-Eclampsia as Underlying Cause for Perinatal Deaths: Time for Action." *Global Health: Science and Practice* 3, no. 4 (2015): 525–527.

12. Sibai, Baha, et al. "Pre-Eclampsia." *The Lancet* 365, no. 9461 (2005): 785–799.

13. Costigan, Kathleen A., et al. "Pregnancy Folklore Revisited: The Case of Heartburn and Hair." *Birth* 33, no. 4 (2006): 311–314.

14. Hildingsson, Ingegerd, et al. "How Long Is a Normal Labor? Contemporary Patterns of Labor and Birth in a Low-Risk Sample of 1,612 Women from Four Nordic Countries." *Birth* 42, no. 4 (2015): 346–353.

15. Petersen, Emily E., et al. "Racial/Ethnic Disparities in Pregnancy-Related Deaths—United States, 2007–2016." *Morbidity and Mortality Weekly Report* 68, no. 35 (2019): 762.

16. Memon, Hafsa, et al. "Pelvic Floor Disorders Following Vaginal or Cesarean Delivery." *Current Opinion in Obstetrics & Gynecology* 24, no. 5 (2012): 349–354.

17. Nygaard, Ingrid, et al. "Prevalence of Symptomatic Pelvic Floor Disorders in US Women." *Journal of the American Medical Association* 300, no. 11 (2008): 1311–1316.

18. Milsom, Ian, et al. "Epidemiology of Urinary Incontinence (UI) and Other Lower Urinary Tract Symptoms (LUTS), Pelvic Organ Prolapse (Pop) and Anal Incontinence (AI)." In *Incontinence: 5th International Consultation on Incontinence,* edited by Paul Abrams, Linda Cardozo, Saad Khoury, and Alan J. Wein, 15–107. Paris: ICUD-EAU, 2013.

19. Neu, Josef, et al. "Cesarean versus Vaginal Delivery: Long-Term Infant Outcomes and the Hygiene Hypothesis." *Clinics in Perinatology* 38, no. 2 (2011): 321–331.

20. Dominguez-Bello, et al. "Partial Restoration of the Microbiota of Cesarean-Born Infants via Vaginal Microbial Transfer." *Nature Medicine* 22, no. 3 (2016): 250–253.

21. DeSilva, Jeremy M., et al. "Neonatal Shoulder Width Suggests a Semirotational, Oblique Birth Mechanism in *Australopithecus afarensis*." *The Anatomical Record* 300, no. 5 (2017): 890–899.

22. Rosenberg, Karen R., et al. "An Anthropological Perspective on the Evolutionary Context of Preeclampsia in Humans." *Journal of Reproductive Immunology* 76, no. 1–2 (2007): 91–97.

23. Hirata, Satoshi, et al. "Mechanism of Birth in Chimpanzees: Humans Are Not Unique among Primates." *Biology Letters* 7, no. 5 (2011): 686–688.

24. Demuru, Elisa, et al. "Is Birth Attendance a Uniquely Human Feature? New Evidence Suggests That Bonobo Females Protect and Support the Parturient." *Evolution and Human Behavior* 39, no. 5 (2018): 502–510.

25. Young, Sharon M., et al. "The Conspicuous Absence of Placenta Consumption in Human Postpartum Females: The Fire Hypothesis." *Ecology of Food and Nutrition* 51, no. 3 (2012): 198–217.

26. Dunsworth, Holly M. "There Is No 'Obstetrical Dilemma': Towards a Braver Medicine with Fewer Childbirth Interventions." *Perspectives in Biology and Medicine* 61, no. 2 (2018): 249–263.

27. Dunsworth, Holly M., et al. "Metabolic Hypothesis for Human Altriciality." *Proceedings of the National Academy of Sciences* 109, no. 38 (2012): 15212–15216.

28. Ziemer, Mary M., et al. "Methods to Prevent and Manage Nipple Pain in Breastfeeding Women." *Western Journal of Nursing Research* 12, no. 6 (1990): 732–744.

29. Vieira Borba, Vânia, et al. "Breastfeeding and Autoimmunity: Programing Health from the Beginning." *American Journal of Reproductive Immunology* 79, no. 1 (2018): e12778. doi:10.1111/aji.12778

CONCLUSION: A Whale in the Water

1. Van de Waal, Erica, et al. "Potent Social Learning and Conformity Shape a Wild Primate's Foraging Decisions." *Science* 340, no. 6131 (2013): 483–485.

2. Allen, Jenny, et al. "Network-based Diffusion Analysis Reveals Cultural Transmission of Lobtail Feeding in Humpback Whales." *Science* 340, no. 6131 (2013): 485–488.

3. Olkowicz, Seweryn, et al. "Birds Have Primate-Like Numbers of Neurons in the Forebrain." *Proceedings of the National Academy of Sciences* 113, no. 26 (2016): 7255–7260.

4. Kochiyama, Takanori, et al. "Reconstructing the Neanderthal Brain Using Computational Anatomy." *Scientific Reports* 8, no. 1 (2018): 6296.

5. Marino, Lori, et al. "Origin and Evolution of Large Brains in Toothed Whales." *The Anatomical Record Part A: Discoveries in Molecular, Cellular, and Evolutionary Biology: An Official Publication of the American Association of Anatomists* 281, no. 2 (2004): 1247–1255.

Index